D0222261

THE EVOLUTION OF INDIVIDUALITY

THE EVOLUTION OF INDIVIDUALITY

LEO W. BUSS

PRINCETON UNIVERSITY PRESS

PRINCETON, NEW JERSEY

Copyright © 1987 by Princeton University Press
Published by Princeton University Press,
41 William Street, Princeton, New Jersey 08540
In the United Kingdom: Princeton University Press,
Guildford, Surrey
All Rights Reserved
Library of Congress Cataloging in Publication Data will be
found on the last printed page of this book
ISBN 0-691-08468-8
     0-691-08469-6 (pbk.)
This book has been composed in Garamond
Clothbound editions of Princeton University Press books
are printed on acid-free paper, and binding materials are
chosen for strength and durability. Paperbacks, although
satisfactory for personal collections, are not usually
suitable for library rebinding.
Printed in the United States of America by
Princeton University Press,
Princeton, New Jersey

CONTENTS

PREFACE

Holistic approaches to biology, as practiced by ecologists, paleontologists, and evolutionists, have traditionally been at odds with reductionist approaches to biology, as practiced by molecular biologists and biochemists. While I whole-heartedly embrace John Tyler Bonner's sentiment—"What is utterly baffling to me is why one cannot be a reductionist and a holist at the same time"[1]—one rather ugly fact remains: evolutionary biologists have only rarely been able to make specific predictions regarding the patterns studied by the reductionist research tradition. Nor have the discoveries of reductionists, until quite recently, led to any obvious resolution of the traditional subject matter of evolutionists.

However painful this observation, it is a fair epitome of biology in the twentieth century. The condition is a curious one. A naive observer of the biological sciences would quickly discover that virtually all biologists are quite content with the proposition that the whole of biological diversity is attributable to evolutionary processes. Yet, that same observer might well be perplexed to learn that, for all the respect the theory holds, entire disciplines of biology seek little or no guidance from evolutionary theory in their everyday empirical work.

Happily, this landscape is rapidly changing. Molecular biology has suddenly become a comparative, and inevitably evolutionary, discipline. A new "fossil record," writ in the genome, is now accessible and is being read in a necessarily piecemeal fashion. The breakthroughs have been technological rather than conceptual. In the context of the revolution, evolutionary theory has remained relatively passive—no more obviously equipped to predict the details of ontogeny, the metabolic complexity of cell architecture, or the fascinating fluidity of the genome than when the synthetic theory of evolution was framed six decades ago.

In this text I advocate a modification of the synthetic the-

1. Bonner, J. T. *The Evolution of Complexity*. Princeton University Press, 1988:4.

ory of evolution which I believe holds the potential for specific evolutionary predictions regarding the natural history
of development, cell structure, and genomic organization.
At the heart of my arguments is the simple observation that
the history of life is a history of the elaboration of new self-
replicating entities by the self-replicating entities contained
within them (or the incorporation of some self-replicating
entities by others). Self-replicating molecules created self-
replicating complexes, such complexes created (or became
incorporated into) cells, cells obtained organelles, and cellular complexes gave rise to multicellular individuals. At
each transition—at each stage in the history of life in which
a new self-replicating unit arose—the rules regarding the
operation of natural selection changed utterly.

The history of life is a history of different units of selection.[2] Novel selective scenarios dominate at times of transition between units of selection. Whereas the lower self-replicating unit was previously selected by the external
environment alone, following the transition it became selected by traits expressed by the higher unit. Variants expressed in the lower unit influence not only the relative replication rate of the lower unit, but also that of the higher
unit. The potential clearly exists for variants to have a synergistic effect (that is, to favor the replication of both the
lower and the higher unit), or for conflicts to arise. The organization of any unit will come to reflect those synergisms
between selection at the higher and the lower levels which
permit the new unit to exploit new environments and those
mechanisms which act to limit subsequent conflicts between
the two units. This explicitly hierarchical perspective on evolution predicts that the myriad complexities of ontogeny,
cell biology, and molecular genetics are ultimately penetrable in the context of an interplay of synergisms and conflicts
between different units of selection.

The view advocated here is but a simple extension of our

2. The phrase "unit of selection" has been subsumed, in the usage of some
philosophers and biologists, by the term "individuality" (Ghiselin, M. T.
1974. *System. Zool.* 25:536–544; Hull, D. L. 1976. *Syst. Zool.* 25:174–
191; Hull, D. L. 1980. *Ann. Rev. Ecol. Syst.* 11:311–332). I have chosen
here not to adopt this convention. Rather, I use the term "individual" in
its more commonplace meaning. An individual is a physiologically discrete organism.

current framework and is not, in and of itself, novel. Wilhelm Roux advocated a hierarchical perspective on evolution nearly a century ago. Indeed, hierarchical perspectives on evolution are undergoing a rebirth among paleobiologists at the moment. Such advances as may be found in this series of essays lie in the *application* of a hierarchical perspective on evolution. I have concentrated on one transition in the history of life, that between the cell and the multicellular individual. I interpret the mechanisms of eukaryotic development to be its evolutionary byproduct.

The text is presented as a series of essays, written in the spirit of G. C. Williams's observation that "one of the strengths of scientific inquiry is that it can progress with any mixture of empiricism, intuition, and formal theory that suits the convenience of the investigator."[3] Each chapter is preceded by a summary. In the first section, I explore the genesis of the modern preoccupation with the individual as the primary unit of selection, tracing it to the ideas of August Weismann and illustrating how Weismann's theories made it unlikely that the synthetic theory of evolution would predict patterns in ontogeny. In the second and third chapters I elaborate a theory of metazoan development. The conservatism of early ontogeny, reflected in patterns of cleavage, gastrulation, and maternal predestination, is argued to reflect conflicts and synergisms between selection at the level of the individual and selection at the level of the cell lineage in the face of a common ancestral constraint on cell division. The diversity of late ontogeny is interpreted in a similar fashion: the manifestly epigenetic character of ontogeny is interpreted as reflecting ancient synergisms between the two units of selection, the establishment of metazoan "bauplans" interpreted as fixation of controls on potential conflicts, and the establishment of heterochrony as the principal vehicle for evolutionary change as the consequence of phylogenetic shifts in the effectiveness of these controls. In the fourth chapter, I explore the evolution of plant and fungal development against the backdrop developed for animals, and interpret the phylogenetic distribution of life cycle traits in terms of the exploitation of synergisms and

3. Williams, G. C. *Adaptation and Natural Selection*. Princeton University Press, 1966:20.

mediation of conflicts between those units of selection which alternate in modern life cycles. The final chapter asks why life has become hierarchically organized, by applying the perspective presented for the evolution of development to other transitions in the history of life.

ACKNOWLEDGMENTS

This book is a product of gifts received from three individuals. Jeremy B. C. Jackson taught me the art of comparative biology. He challenged me to view the seemingly endless morass of specific detail in the literature not as justification for narrowing one's scope of inquiry, but as fodder upon which any truly general idea will ravenously feed. John T. Bonner gave me what he has given so many others: a sense of his own fascination for development. Doug R. Green always seemed to understand what I was thinking about and convinced me that I could say it.

Special thanks must go to the members of my laboratory for enduring the many absences the writing of this book necessitated and the surely no less difficult periods of my presence. Particular thanks go to Philip Yund and Jan Taschner. My friends Mark Bertness and Doug Green knew when my enthusiasm was waning and, as only close friends can, applied just the appropriate stimuli to renew my excitement with the project. Discussions with Philip Yund led to calculations on the limits of life cycle evolution. J. Rimas Vaisnys asked penetrating questions which considerably clarified the discussion of the evolution of hierarchical organization. The manuscript has been read in full by Neil Blackstone, Mark Bertness, Jane Moore Buss, John Bonner, Cliff Cunningham, Doug Green, G. Evelyn Hutchinson, Lynn Rothschild, Keith Thomson, J. Rimas Vaisnys, Geerat Vermeij, Elizabeth Vrba, and Philip Yund. Any errors remaining are not entirely their fault.

Support was generously provided by a fellowship from the John Simon Guggenheim Memorial Foundation, for which I am most grateful. Thanks also go to Yale University for their liberal leave policy for junior faculty members and to the National Science Foundation for their continued support of my empirical research.

Several "removal experiments," forced by our respective careers, separated my wife and me during the gestation of

xii Acknowledgments

the ideas presented here. Not a single page written in her absence found its way into the final text. The absence of appropriate controls notwithstanding, I confidently conclude that this book could not have been completed without her. This work is in celebration of, and is dedicated to, Jane Moore.

FIGURE SOURCES

1.1   Simpson, G. G., C. S. Pittendrigh, and L. H. Tiffany. *Life: An Introduction to Biology*. New York: Harcourt, Brace, and Co., 1957. Reprinted by permission.

1.2   Slack, J.M.W. *From Egg to Embryo*. Cambridge University Press, 1983.
        Raff, R. A. and T. C. Kaufman. *Embryos, Genes, and Evolution*. New York: Macmillan, 1983.

1.3   Weismann, A. *The Evolution Theory*. Vol. 1. London: Edward Arnold, 1904.

2.1   Liesche, W. 1938. *Arch. Protistenk*. 91:135–186.

2.2   Cleveland, L. R. 1938. *Biol. Bull*. 74:41–55.
        Cleveland, L. R. in L. Levine, ed. *The Cell in Mitosis*. New York: Academic Press, 1971.

2.3   Grell, K. G. *Protozoology*. Berlin: Springer-Verlag, 1973. Reprinted by permission.

2.4   Belar, K. 1922. *Arch. Protistenk*. 46:1–96.

2.5   Saville Kent, W. *A Manual of the Infusoria*. Vol. 3. London: David Bogue, 1882.

2.6   Beklemishev, W. N. *Principles of Comparative Anatomy of Invertebrates*. Vol. 1. Trans. J. M. MacLenman, ed. Z. Kabata. University of Chicago Press, 1969. Reprinted by permission.

2.7   Minchin, E. A. 1896. *Proc. Roy. Soc. Lond*. 60:42–52.

2.8   Holtfreter, J. 1943. *J. exp. Zool*. 93:251–323.

2.9   Claus, C. *Untersuchungen über die Organisation und Entwicklung der Medusen*. Prag und Leipzig, 1883.

2.10   Minchin, E. A. in E. R. Lankester, ed. *A Treatise on Zoology*. London: Adam and Charles Black, 1900.

2.11   Harm, K. 1903. *Z. Wiss. Zool*. 73:115–165.

2.12   Balfour, F. M. *A Treatise on Comparative Embryology*. Vol. II. London: Macmillan, 1881.

2.13   Iijima, I. 1884. *Z. Wiss. Zool*. 40:359–404.
        Mattiesen, E. 1904. *Z. Wiss. Zool*. 77:274–361.

2.14   Claus, C. *Untersuchungen über die Organisation und Entwicklung der Medusen*. Prag und Leipzig, 1883.

2.16   Dubosocq, O. and O. Tuzet. 1942. *Arch. Zool. Exp. Gén.* 82:151–163.

2.17   Hörstadius, S. 1939. *Biol. Rev.* 14:132–179.

2.18   Kume, M. and K. Dan. *Invertebrate Embryology*. Tokyo: Bai Fu Kan Press, 1957. Reprinted by permission.

2.19   Kato, K. 1940. *Jap. J. Zool.* 8:251–254.

2.20   Costello, D. P. 1945. *J. exp. Zool.* 100:19–66.

2.21   Raff, R. A. and T. C. Kaufman. *Embryos, Genes, and Evolution*. New York: Macmillan, 1983.
        Clement, A. C. 1952. *J. exp. Zool.* 121:593–626.

2.22   Titlebaum, A. 1928. *Proc. Nat. Acad. Sci. USA* 14:245–247.

2.23   Nieuwkoop, P. D. and L. A. Sutasurya. *Primordial Germ Cells in the Invertebrates*. Cambridge University Press, 1981.

3.1   Olive, L. S. *The Mycetozoans*. New York: Academic Press, 1975.
        Bonner, J. T. *The Cellular Slime Molds*. 2nd ed., Princeton University Press, 1967.

3.2   Bonner, J. T. *Ibid*.

3.3   Weismann, A. *The Evolution Theory*. Vol. 1. London: Edward Arnold, 1904.

3.4   Tuzet, O. in E. Wolf, ed. *L'Origine de la Lignée Germinale chez les Vertébrés et chez quelques Groupes d'Invertébrés*. Paris: Hermann, 1964.

3.5   Bonner, J. T. *The Evolution of Development*. Cambridge University Press, 1958.

3.6   Margulis, L. *Early Life*. New York: Van Nostrand Reinhold, 1982. Reprinted by permission.

3.7   Marrack, P. and J. Kappler, 1986. *Sci. Amer.* 254:36–54.

3.9   Conway Morris, S. 1977. *Palaeontology* 20:623–40.

3.10   Hörstadius, S. 1935. *Pubbl. Staz. Zool. Napoli.* 14:251–429.

3.11   Ede, D. A. *An Introduction to Developmental Biology*. New York: Wiley, 1978.

3.12   Coulombre, A. J. in R. L. Dettaan and H. Ursprung, eds. *Organogenesis*. New York: Holt, Rinehart, and Winston, 1965. Reprinted by permission.

3.13   Raff, R. A. and T. C. Kaufman. *Embryos, Genes, and Evolution*. New York: Macmillan, 1983.

3.15   MacBride, E. W. *Textbook of Embryology*. I. *Invertebrata*. London: Macmillan, 1914.
    Wilson, E. B. 1904. *J. exp. Zool.* 1:197–268.

3.16   Lillie, F. R. 1895. *J. Morphol.* 10:1–100.

3.17   Douce, M. *Histoire Naturelle des Annelés*. Paris: Librairie Encyclopédique de Roet, 1865.

3.18   Raff, R. A. and T. C. Kaufman. *Embryos, Genes, and Evolution*. New York: Macmillan, 1983.

4.1   Burnett, J. H. *Fundamentals of Mycology*. 2nd ed. London: Edward Arnold, 1976. Reprinted by permission.

4.2   Bracker, C. E. and E. E. Butler. 1963. *Mycologia* 55:35–38. Reprinted by permission.

4.3   Moore-Landecker, E. *Fundamentals of the Fungi*. Englewood Cliffs, New Jersey: Prentice-Hall, 1972. Reprinted by permission.

4.4   Garrone, R. 1974. *Arch. Anat. microsc. Morph. exp.* 63:163–182. Reprinted by permission.

4.5   Olsen, O. W. *Animal Parasites*. Baltimore: University Park Press, 1974. Reprinted by permission.

4.6   Muller, W. 1964. *Wilhelm Roux' Arch. Entwicklungsmech.* 155:181–268.

4.7   Buller, A.H.R. *Research on Fungi*. Vol. 4. London: Longmans, Green, and Co., 1931.

4.8   Raper, J. R. in E. G. Butler, ed. *Biological Specificity and Growth*. Princeton University Press, 1955.

4.9   Burnett, J. H. *Fundamentals of Mycology*. 2nd ed. London: Edward Arnold, 1976. Reprinted by permission.

4.10   Taylor, W. R. *Marine Algae of the Northeastern Coast of North America*. University of Michigan Press, 1937.
    Bold, H. C. *Morphology of Plants*. 2nd ed. New York: Harper and Row, 1967. Used by permission.

4.11   Bold, H. C. *Morphology of Plants*. 2nd ed. New York: Harper and Row, 1967. Reprinted by permission.

AUGUST WEISMANN'S LEGACY

There are books in plenty on experimental embryology but none on theoretical embryology. Why is this?—J. H. WOODGER, 1930

Summary

The late-nineteenth-century naturalist August Weismann framed the modern biological view of the individual with his "Doctrine of the Continuity of the Germ Plasm." Weismann's great contribution was the recognition that heritability is controlled by development. Terminal somatic differentiation denies a cell lineage the opportunity to contribute to subsequent generations. Weismann's doctrine justified the axiom that evolutionary change can be reduced to a consequence of selection between individuals within populations. If variation arising in the course of ontogeny is not heritable, the dynamics of cell lineages within the somatic environment are a matter of little direct evolutionary interest.

Weismann's doctrine was merely one prediction of his wide-ranging theories of inheritance. He believed that only germ cells contained the heritable material and that somatic cells adopted their supportive roles as a consequence of an irreversible conversion of the "germ plasm" to differentiated structures. In the context of his inheritance theories, the doctrine was unassailable. Somatic cells could not be heritable because their heritable material had been converted into the products of differentiation. While Weismann's inheritance theories were ultimately proved fictional, their corollary, that the individual is the sole unit of biological organization, was nevertheless incorporated as a tacit assumption in the modern synthetic theory of evolution.

Patterns of development which violate Weismann's doctrine were well known at the time when Mendelian genetics became wedded to natural history to form the modern synthetic theory of evolution. Yet the Modern Synthesis was an intellectual event virtually unattended by embryologists. A curious lack of communication between embryologists and geneticists as the synthetic theory was being developed—itself a consequence of differences of opinion on Weismann's inheritance theories—prevented a broadening of the theory to accommodate the diversity of ontogenetic pattern. The geneticists and naturalists who authored the Modern Synthesis had no pressing reason to raise embryological concerns themselves, as all worked on organisms in which the Weismannian ideal of the individual was closely approx-

imated. A synthetic theory came into being which was and is at variance with known developmental patterns.

Weismann's doctrine would be justified, despite its flawed origins, if terminal determination of the germ line always occurred in earliest ontogeny. However, taxa differ in their mode of development. In some taxa, this Weismannian assumption is closely approximated; in others it is not. Crucially, the phyletic distribution of this trait illustrates that early terminal differentiation is a character limited exclusively to some higher metazoan taxa. When multicellular, cellular-differentiated life first arose, Weismann's doctrine was violated. At this point—and in many taxa even today—it is inappropriate to assume that the individual is the sole unit of selection. Individuality is a derived character.

The synthetic theory has enjoyed a half-century of unparalleled success. Its success is no paradox. Rather, acceptance of the Modern Synthesis stands as compelling testimony to the fact that evolution has manifestly favored ontogenies in which the Weismannian ideal is approximated. Yet a theory of evolution which assumes the evolution of individuality as a basal tenet cannot be expected to explain how individuality evolved. A theory of ontogeny has not been framed, and cannot be framed, within the confines of a notion of the individual that is at variance with developmental fact.

I

The Darwinian notion of evolution as a process directed by selection upon heritable variation has not been effectively challenged since Darwin first articulated it. Evolution by common descent had become established as undeniable fact within decades of the publication of the *Origin of Species*. Yet, the original theory lacked any mechanism by which variation might be inherited. Nor did it provide any assurance that such a mechanism, if discovered, could also direct the myriad details of ontogeny. Biologists, content that evolution had occurred, were understandably far from any agreement on *how* it had occurred. Turmoil marked the late nineteenth and early twentieth centuries.

Relative to its early trials, the past four decades have been a remarkably tranquil time for the theory of evolution. The framing of a consistent mechanistic theory began in the 1920's and grew to fruition in the 1940's, with the wedding of Mendelian genetics to natural history to form the "Modern Synthesis." The synthesis formally codified the evolutionary process as a consequence of variation among individuals within populations. The operation of natural selection upon genetically discrete and unique individuals, rather than being minimally perceived as wholly consistent with observed evolutionary change, was, for the first time, held to be the sole agent of all evolutionary modification.

Recent discussions, arising from opposite extremes of the spectrum of biological sciences, while not questioning that selection upon individuals is a *necessary* component of evolutionary change, have come to question whether the synthesis alone is a *sufficient* explanation of the evolutionary process. Paleontologists, having long attributed the lack of continuous directional phenotypic variation to imperfections in the fossil record, have come to ask whether the record might not be more representative than has traditionally been supposed. Perhaps the lack of gradual transformations in fossils should be accepted at face value, implying that evolution has occurred by a process of morphological stasis punctuated by episodic change.[1] If so, species may act as a unit, originating via speciation and being selected on the basis of differential rates of speciation and extinction.[2] Molecular biologists, once content that the central dogma accorded fully with the synthetic theory, are revealing the genome as a surprisingly dynamic entity. Enzymes are described daily by which gene sequences are cut, spliced, digested, rearranged, mutated, reiterated, edited, corrected, inverted, deleted, truncated, and translocated.[3] A surprising degree of autonomy is now conceivable; a gene sequence coding for its own gene-processing enzymes might itself qualify as a unit of selection.[4]

The period of relative tranquility was purchased at the cost of a restriction on the breadth of Darwin's original theory. The logical structure of Darwin's argument allows any

1. Eldredge, N. and S. J. Gould. In T.J.M. Schopf, ed. *Models in Paleobiology*. San Francisco: Freeman, 1972:82–115.
2. Stanley, S. M. 1975. *Proc. Nat. Acad. Sci. USA* 72:646–650.
3. This splendid characterization of the current aura of excitement in molecular biology is paraphrased from a recent article on developmental constraints in evolution (Maynard Smith, J., R. Burian, S. Kauffman, P. Alberch, J. Campbell, B. Goodwin, R. Lande, D. Raup, and L. Wolpert. 1985. *Q. Rev. Biol.* 60:265–287).
4. Doolittle, W. F. and C. Sapienza. 1980. *Nature* 284:601–603; Orgel, L. E. and F.H.C. Crick. 1980. *Nature* 284:604–607.

unit to evolve if it replicates with high fidelity and if selection distinguishes between the variants. Species, populations, and lineages of individuals, cells, organelles, and gene sequences can all potentially evolve.[5] Yet we have been largely content to attribute the whole of biological diversity to selection upon individuals. The once comfortable cloak of the Modern Synthesis has become restrictive.

II

To a considerable extent the modern view of the individual can be traced to the work of August Weismann. Weismann was one of the many theoreticians of the late nineteenth century attempting to forge a coherent theory merging cytology, embryology, and evolution. The period was one in which Darwin's great synthesis was broadly accepted in principle, but widely confused as to mechanism. Naturalists, following Darwin's tradition, found at every turn manifest evidence of evolution by common descent. Anatomists found in comparison a basis for prediction. Embryologists found phylogeny in ontogeny. Yet cytology was still a theoretical science, Mendel's observations lay dormant, and acquired characteristics were widely believed to be heritable.

Weismann's greatest contribution, and the reason for our present concern, was his doctrine of the continuity of the germ plasm. Weismann held that there existed a "molecular distinction" between the germ plasm and the soma, such that the germ plasm represented the immortal link between generations, whereas the soma was merely a mortal vessel upon which selection acts. Heritability was solely the province of the germ line. Weismann's doctrine, though not falsifiable given nineteenth-century technology, was broadly in accord with several classes of available evidence. There was little doubt, for example, that, in many animals, cells destined to become the gametes are sequestered in earliest ontogeny. Nor could one doubt the failure of somatic tissues to reconstitute a new individual following injury. Far more important, though, was the fact that the doctrine denied the possibility of any transmission of acquired characteristics. Weismann's doctrine labels the transmission of acquired characteristics a theoretical impossibility. If the germ plasm was separated from the soma by some molecular barrier, how

5. Lewontin, R. C. 1970. *Ann. Rev. Ecol. Syst.* 1:1–18.

could environmental influences acting on the soma be transmitted to the germ plasm? They could not.

Weismann was the first major figure to demand an uncompromising selectionist stance and to defend this position against all claims of soft inheritance.[6] Accordingly, modern authors often view Weismann's doctrine largely in terms of its impact on Lamarckian thought. Mayr, for example, states that

> We now know that Weismann's basic idea—a complete separation of the germ-plasm from its expression in the phenotype of the body—was absolutely correct. His intuition to postulate such a separation was faultless. However, among the two possible ways for effecting this he selected the separation of the germ cells from the body cells, while we now know that the crucial separation is that between the DNA program of the nucleus and the proteins in the cytoplasm of each cell.[7]

Acquired modifications cannot be inherited because the path from DNA to proteins is a one-way street; changes in DNA may produce changes in proteins, but changes in proteins cannot impress these changes upon DNA.

Weismann, however, saw the doctrine of the continuity of the germ line as something more than an argument about the transmission of acquired characteristics. He clearly states:

> This doctrine is either right or wrong, and there is no middle course: to this extent I quite admit that I am prejudiced. But the question as to whether acquired characters can be impressed upon the germ and thus transmitted would not be by any means settled in this way. . . . It seems to me that the problem of the transmission or non-transmission of acquired characteristics remains, whether the theory of the continuity of the germ-plasm be accepted or rejected.[8]

6. Ernst Mayr (*The Growth of Biological Thought*. Cambridge, Mass.: Harvard University Press, 1982) distinguishes between soft and hard inheritance theories on the basis of whether the view permits the inheritance of environmental influences.

7. *Ibid.*, p. 700.

8. Weismann, A. *Essays on Heredity and Kindred Biological Problems*. E. B. Poulton, S. Schonland, and A. E. Shipley, eds. and trans. Oxford: Clarendon Press, 1889:403.

Weismann's prescience in the matter of acquired characteristics is certainly remarkable. However, it would be a mis-
reading of his doctrine of the continuity of the germ plasm
to hold this to be the sole legacy of the doctrine to modern
biology.

Weismann's doctrine was part of a more far-reaching theory of inheritance. Like most theorists of the time, Weismann was faced with the dilemma of interpreting the fact
that while the cells of most organisms retain the capacity for
cell division, only some cells actually participate in the production of new individuals. He interpreted this fact as evidence that the cells comprising an organism are endowed
with differing molecular determinants. Weismann saw the
nucleus as containing multiple particles which were distributed unevenly at cell division. These particles were *irreversibly converted* into the structures of the developing organism,
causing differentiation. One set of particles, the germ
plasm, did not participate in this process. Rather than causing differentiation by conversion to differentiated cell products, the germ plasm remained inviolate and endowed the
cell containing it with the capacity to generate a new individual. The germ plasm was the heritable material, and only
some cells retained this essential material. Today we know
this to be false. All cells have a full complement of DNA,
albeit differential in their zones of synthetic activities.[9]

Weismann's doctrine of the continuity of the germ line is
a statement, the first clear statement, on the units of selection. The very fact that Weismann stated his theory in terms
of cell lines which differ in the possession of a molecular con-
stitution conferring or denying heritability means that acceptance of the doctrine justifies the view of the individual
as a genetically homogeneous unit. If not all cells contain
heritable material, selection is necessarily on the individual,
not on the cells or their constituents.

9. Weismann may ultimately prove correct on this point in certain restricted circumstances. If genic rearrangement, as is known to occur in the
construction of various receptor molecules of the vertebrate immune system, proves a common mechanism of terminal somatic differentiation,
then Weismann's inheritance theories will be vindicated, albeit in a curious and unsuspected sense. The pronounced chromatin diminution displayed in the somatic lineages of some dipterans and nematodes (which
have not, as yet, been implicated as determinants of somatic cell fate) may
ultimately prove to be another mechanism whereby Weismann's inheritance theories could yet prove viable.

Weismann's doctrine of the continuity of the germ plasm, with its emphasis on the individual as the selective unit and its strict reliance on hard inheritance, changed the landscape of evolutionary biology. Romanes found it necessary to coin the term "Neo-Darwinian" to differentiate pre- and post-Weismannian concepts of evolution.[10] Correns, one of the co-rediscoverers of Mendel's laws, said that the rediscovery of Mendelian rules in 1900 was no great intellectual breakthrough after Weismann had paved the way. Mayr, in his monographic text, *The Growth of Biological Thought*, refers to Weismann as "one of the greatest biologists of all time."[11]

Only two decades after the publication of Weismann's *Der Kiemplasma*, though, Thomas Hunt Morgan referred to Weismann, with derision, as "that philosopher of Freiburg."[12] During the 1890's and continuing through the early twentieth century, the trend was changing from one of speculation to experimentation. The speculative orgy of the late nineteenth century provided a comprehensive evolutionary theory uniting cytology, embryology, and evolution. Although most of this theory, including Weismann's theories of inheritance, ultimately proved fictional, the outcome was the finest result to which theory may aspire: the right questions were asked. Weismann's theories, as those of many others, were based on assumptions as to the mechanisms of inheritance and the developmental potentialities of cells within multicellular organisms. These two issues proved to be the focus of experimentalists for several decades to come. This subsequent work, however, was not directed toward synthesis as the immediate goal, but rather directed toward the accumulation of experimental fact and the careful development of provisional validity.

Shortly after Weismann's synthesis, Mendel's rules were rediscovered and work on transmission genetics proceeded at an unparalleled pace, both in terms of quantifying results of breeding experiments and in terms of interpreting these results in light of cytological findings of the late nineteenth century. Of the many laboratories investigating problems of

10. Romanes, G. J. *Life and Letters*. London: Longmans, Green, 1896.
11. Mayr, E. *Growth of Biological Thought*, p. 698.
12. The degree to which Weismann's concept of the function of theory in biology stood in opposition to that of Morgan's is explored in depth by Garland E. Allen (*Thomas Hunt Morgan: The Man and His Science*. Princeton University Press, 1978).

transmission genetics, Morgan's laboratory stands preeminent. In part this certainly reflects the extraordinary energy and breadth of Morgan's collaborators.[13] More fundamentally, though, the success of Morgan's group was the result of the explicit program of the laboratory to divorce problems of transmission genetics from problems of cell physiology and ontogeny. It was through relentless concentration on the frequency of the expression of traits, without concern for the manner in which traits may be expressed, that transmission genetics grew as a science.

During the same time interval, embryologists turned from phylogenetic speculation to Roux's *Entwicklungsmechanik* (developmental mechanics). Embryologists began to attempt to identify cause-and-effect relationships in ontogeny as never before. This revolution, parallelling that in transmission genetics, was primarily one of cognitive style, characterized by the insistence of its founders on experimentation rather than observation. Embryological thought shifted from Weismannian concentration on heredity and phylogeny to a concern for the processes by which an individual developed. This trend is no more clearly illustrated than by C.O. Whitman's statement:

> We have no longer any use for the 'Ahnengalleries' [ancestor portrait gallery] of phylogeny. . . . We are no better off for knowing that we have eyes because our ancestors had eyes. If our eyes resemble theirs it is not on account of genealogical connection, but because the molecular germinal basis is developed under similar conditions.[14]

Experimental embryology flourished, but flourished in isolation from both studies in transmission genetics and natural history.

13. Historians often understate Morgan's role relative to that of his collaborators. While the impact of Bridges, Muller, and Sturtevant on the development of transmission genetics as a science cannot be overemphasized, neither should Morgan's role as the catalyst be forgotten. A. H. Sturtevant (1965. *Proc. Am. Phil. Soc.* 109:199–204) testifies: "There have been few times and places in scientific laboratories with such an atmosphere of excitement and with such a record of sustained enthusiasm. This was due in large part to Morgan's own attitude, compounded of enthusiasm combined with a strong critical sense, generosity, open-mindedness, and a remarkable sense of humor."

14. Whitman, C. O. *Biological Lectures: The Marine Biological Laboratory of Woods Hole, Mass.* Boston: Ginn and Co., 1895:iii–vii.

Viewing the past with the arrogance of the present, one might think that like-minded experimentalists in embryology and transmission genetics would have found fertile ground for synthesis. However, it must be remembered that embryologists, concerned with differentiation, had difficulty accepting Mendelian genetics. The rejection is clearly stated by F. R. Lillie:

> The present postulate of genetics is that the genes are always the same in a given individual, in whatever place, at whatever time, within the life-history of the individual, except that the occurrence of mutations or abnormal disjunctions, to which the same principles then apply. The essential problem of development is precisely that differentiation in relation to space and time within the life-history of an individual which genetics appears implicitly to ignore. The progress of genetics and physiology of development can only result in sharper definition of these two fields, and any expectation of their reunion (in a Weismannian sense) is in my opinion doomed to disappointment.[15]

As Lillie predicted, there was not, nor could there be, a synthesis of embryology and genetics in the context of Weismann's doctrine. Experimental embryologists, despite enormous success in defining the varied consequences of manipulation of eggs and embryos, were far less successful in attempts to provide genetic interpretations for these phenomena.

The result is a curious historical confusion. Weismann's inheritance theories predicted that somatic tissues were not heritable and that acquired modifications could not be inherited. Geneticists quickly discarded Weismann's inheritance theories, while accepting his predictions regarding soft inheritance. Embryologists, in contrast, remained undecided on his inheritance theories—as his theories explicitly addressed the central issue of differentiation in space and time through ontogeny—while they had ample reasons to doubt his predictions. It is apparent that the experimental traditions in embryology and in genetics were not equivalently successful in resolving the issues formulated by the speculative era they supplanted. In retrospect, it seems evident that the problems of developmental genetics could not

15. Lillie, F. R. 1927. *Science* 66:361–368.

have been explored without an understanding of molecular problems that were not even broached until the 1950's. Although the data of geneticists ultimately forced embryologists from the Weismannian view of inheritance, this difficulty separated embryologists from geneticists during the period in which both matured as experimental sciences.[16]

The reluctance of embryologists to embrace Mendelian genetics had a profound effect on the development of evolutionary theory. The celebrated Modern Synthesis, through which Mendelian genetics was wedded to systematics, was an event in which embryologists played virtually no role. The difficulties which originally separated embryologists and geneticists were issues left unaddressed. The Modern Synthesis codified Mendelian genetics in terms of populations of individuals. Explicit in these formulations was the assumption that individuals could be treated as genetically homogeneous units. John Maynard Smith clearly articulates the assumption and its source in Weismannian thought, noting that:

> After the publication of Darwin's *Origin of Species*, but before the general acceptance of Weismann's views, problems of evolution and development were inextricably bound up with one another. One consequence of Weismann's concept of the separation of the germ line and soma was to make it possible to understand genetics, and hence evolution, without understanding development. In the short run this was an immensely valuable contribution, because the problems of heredity proved to be soluble, whereas those of development

16. The extent to which the field of genetics was at odds with embryology is reflected in an anecdote recounted by Viktor Hamburger in his commentary on the minimal impact of embryological thought in the framing of the Modern Synthesis (in E. Mayr and W. B. Provine, eds. *The Evolutionary Synthesis: Perspectives on the Unification of Biology.* Cambridge, Mass.: Harvard University Press, 1980:97–112). Hamburger notes that: "Morgan's book *Embryology and Genetics* illustrates the ambiguity of this situation. In its preface, Morgan wrote: 'The story of genetics has been so interwoven with that of experimental embryology that the two can now, to some extent, be told as a single story. . . . It is possible to attempt to weave them together in a single narrative.' The story goes that after the publication of the book, Morgan asked a prominent visitor what he thought of it. The visitor frankly responded that he could not find a synthesis of the two fields; whereupon Morgan, tongue in cheek, asked 'What does the title say?' " (pp. 100–101).

apparently were not. . . . My own view is that development remains one of the most important problems of biology, and that we shall need new concepts before we can understand it. It is comforting, meanwhile, that Weismann was right."[17]

Weismann, however, was right for the wrong reasons. Yet his doctrine became a tacit assumption of the Modern Synthesis. The view of the individual adopted by the synthetic theory is that of Weismann. The individual is held to be a discrete unit, with heritability limited to a very small subset of genetically homogeneous cells. This view is not only derived from an erroneous theory—a theory implying that only some cells contain DNA—but it is also a view which is at considerable variance with the established facts of development.

III

George Gaylord Simpson, one of the principal authors of the Modern Synthesis, schematized the assumption of genetic individuality in an explicit form, labeling it as Weismann's doctrine.[18] Many of us first encountered a version of Simpson's figure in secondary school (Figure 1.1). This now-classic view has the zygote producing somatic cells via mitosis and germ cells via meiosis. Genetic variation arising during the course of ontogeny cannot be inherited. Heritable variation only occurs in the zygote or during the reduction divisions of gametogenesis. This is an appealing diagram. It represents the ideal of the individual as a unique, genetically homogeneous, entity. It is an ideal, however, that is only approximated in real organisms.

Multicellular organisms are composites. Individuals are composed of cells capable of division and of variation. Within eukaryotic cells are organelles, also capable of reproduction and variation, and within organelles and nuclei are gene sequences which may also have these capabilities. As Weismann recognized, *heritability is controlled by development*. A unicellular alga dividing by mitosis to produce a clone of

17. Maynard Smith, J. *Evolution and the Theory of Games*. Cambridge University Press, 1982:6.
18. Simpson, G. G., C. S. Pittendrigh, and L. H. Tiffany. *Life: An Introduction to Biology*. New York: Harcourt, Brace, and Co., 1957:281.

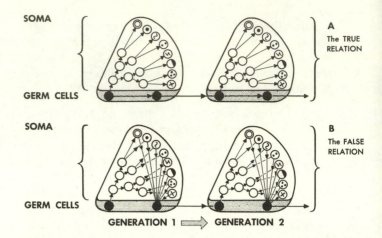

FIGURE 1.1   Schematic diagram of Weismann's doctrine from George Gaylord Simpson's influential textbook *Life: An Introduction to Biology*. Here somatic cells (variously filled circles) arise by mitosis and germ cells (closed circles in shaded region) by meiosis. This view excludes heritability of variants which arise in stem cells (open circles). (From Simpson, Pittendrigh, and Tiffany, 1957.)

daughter cells differs from a metazoan zygote dividing by mitosis likewise to produce a clone of cells in that all descendants of the former are capable of giving rise to a new multicellular individual, whereas only a fraction of the clonemates of the latter retain this ability. Irreversible differentiation of cells to purely somatic function denies a lineage the capacity to generate a new organism. Unless a cell lineage produces a gamete or retains totipotency and succeeds in asexually producing a new organism capable of further propagation, genetic variation within that cell lineage will not be heritable.

Heredity as studied today is a matter of macromolecules, information, and code. However intriguing the newly discovered fluidity of the genome, knowledge of the molecular mechanics of heredity is *not* equivalent to knowledge of the units which prove heritable. Recognition that the processes of development control heritability focuses the issue of inheritance from questions directed purely at the molecular mechanisms of transmission to questions of patterns in developmental determinism. If modes of development are such that only a few cell divisions are intercalated between fertilization and the terminal determination of the germ line each generation, then the opportunity for heritable variation to arise in the course of ontogeny is small and the view of the individual as a genetically homogeneous unit is largely vindicated. If, however, patterns in development allow substantive opportunities for embryonic cells to vary and yet still gain access to the gametes, then genetic variation arising during the ontogeny of an individual must be acknowledged as a potentially important source of transmissible variation.

The fact that development controls heritability may be illustrated by comparing the development of two common laboratory animals. Consider first the development of the dipteran *Drosophila melanogaster* (Figure 1.2). Following fertilization and syngamy, the egg nucleus undergoes thirteen cleavage divisions in rapid sequence. The *Drosophila* egg, like any egg, has two sources of mRNA to draw upon; that drawn from the cytoplasm and provided by the mother, and that synthesized by the embryo itself. In *Drosophila*, the first appreciable mRNA synthesis by the embryo is found only after the thirteenth cleavage division. Development prior to this point is entirely directed by instructions left by the parent.[19] These first thirteen cleavage divisions are not accompanied by cell division; the embryo at this stage is a coenocytic mass. Following the thirteenth nuclear division, however, cell division takes place and the resulting embryo is composed of two distinct regions: the pole cells and the cellular blastoderm. The pole cells are the primordial germ cells and the roughly 6,000 cells comprising the cellular blastoderm are the somatic cells. It is at this point in devel-

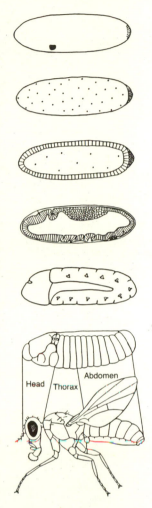

FIGURE 1.2

Stages in the development of *Drosophila melanogaster*. (After Slack, 1983 and Raff and Kaufman, 1983.)

19. Using a wide variety of techniques, several investigators have confirmed that the *Drosophila* embryo displays an exceedingly low rate of RNA synthesis prior to blastoderm formation (e.g., Fausto-Sterling, A., L. M. Zheutlin, and P. R. Brown. 1974. *Dev. Biol.* 40:78–83). That this small, but detectable, RNA synthesis is not fully responsible for germ cell determination is demonstrated by the occurrence of maternal effect mutants acting during early determination (Rice, T. B. and A. Garen. 1975. *Dev. Biol.* 43:277–286). Eggs, for example, bearing one of these, *mat(3)1*, form no somatic blastoderm, but normal pole cells. The extent of maternal control is confirmed by nuclear transplantation experiments which demonstrate that nuclei are totipotent prior to blastoderm cellularization (Illmensee, K. 1972. *Wilhelm Roux' Arch. Entwicklungsmech.* 170:267–298; Okada, M., I. A. Kleinman, and H. A. Schneiderman. 1974. *Dev. Biol.* 39:286–294).

opment that determination of cell fate takes place for larval, as well as adult, structure.

Only thirteen nuclear divisions are intercalated between the zygote and the terminal determination of those cells destined to become the zygotes of the next generation. A variant would have to arise within this brief interval in order to be heritable. But since development up to this point is directed solely by maternal instructions, there is no opportunity during the entire life cycle of *Drosophila* for any cell to influence its own fate by products of its own making. Simpson's scheme is vindicated, as the germ is derived directly from the zygote under maternal instructions, generation after generation.

Contrast *Drosophila* development with that of the simple freshwater hydroid *Hydra* (Figure 1.3). The zygote gives rise to an embryo composed of two distinct populations of cells: the interstitial (or I-cells) and somatic cells. The I-cells are a multipotent cell lineage which may, under the proper stimulus, give rise to any of the various somatic cell types.[20] Some somatic cell types are capable of further cell division and growth, some are not. The latter category of somatic cells must be continuously replenished by differentiation of I-cells. Both classes of somatic cells, however, share an inability to either differentiate into different somatic cell types or to return to the multipotent status of I-cells.

The zygote of a *Hydra* gives rise to a polyp which, under favorable conditions, will reproduce asexually. Asexual reproduction involves the movement of multipotent I-cells and various somatic tissues into a bud off the parent polyp. This bud soon detaches and assumes an independent existence. The asexual reproductive phase may be of indeterminate length; investigators maintain stock asexually for decades. When local conditions deteriorate, *Hydra* may be induced to cease reproduction by asexual iteration, and instead I-cells differentiate into gametes. However, between

---

20. Interstitial cells, recognized by their distinctive amoeboid morphology, have been observed to give rise to various somatic lineages in several hydroids. Gross morphological identity has traditionally formed the basis for the assumption that these cells represent a single multipotent lineage. Hans Bode and others are actively testing the possibility that the morphological identity of interstitial cells may mask an underlying complex of partially differentiated stem cells (as, for example, is the case in lymphocytes).

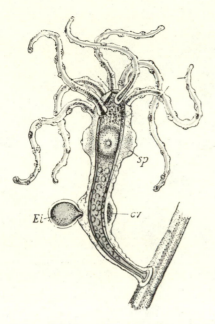

FIGURE 1.3   Polyp of *Hydra viridis*, showing testis (sp), ovary
(ov), and developing embryo (Ei). (From Weismann, 1904.)

each sexual generation an indeterminate number of asexual
iterations may occur. In contrast to *Drosophila*, where totipotency is limited after thirteen nuclear divisions, the I-cells
of *Hydra* remain multipotent and mitotically active
throughout the potentially great life span of the animal.

The *Drosophila* example fits Simpson's figure well, the *Hydra* hardly at all. Simpson's figure may be modified, as in
Figure 1.4, to accommodate hydroids. The modified view
holds that the zygote gives rise, by mitosis, to a totipotent
cell lineage which may have one of three fates: (1) it may give
rise to somatic cells, (2) it may undergo reduction divisions
and give rise to gametes, or (3) it may retain totipotency and
undergo continuing episodes of differentiation into somatic
cells or gametes.

Now let us consider the potential for heritability of suborganismal variation in these two developmental modes.
Any variant arising in the totipotent lineage is in the pool
from which the gametes are drawn. The likelihood that a genetic variant will occur in this pool of cells is a function of
the basal rate at which variation arises per division and the

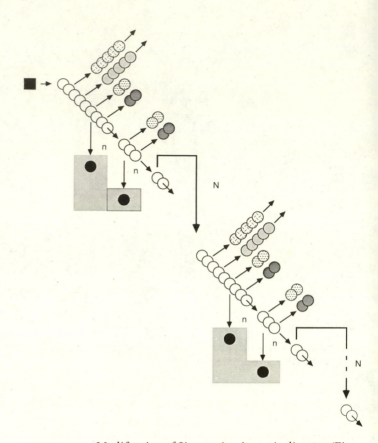

FIGURE 1.4   Modification of Simpson's schematic diagram (Figure 1.1) which recognizes the ability of totipotent stem cells (open circles) to give rise to gametes (closed circles) and recognizes the potential for repeated asexual iteration of new physiologically discrete individuals from a single zygote (closed box). The number of totipotent divisions per sexual generation (n) and the number of asexual generations per sexual generation (N) varies in a taxon-specific fashion and may be very large. This figure represents a conservative modification of Simpson's diagram as it does not permit functional somatic cells (variously filled circles) to contribute to the gametes, a process known to occur in some metazoans (see Figure 3.4).

number of divisions made by the totipotent lineage.[21] In *Drosophila melanogaster*, for example, the totipotent lineage undergoes thirteen nuclear divisions per sexual generation. The potential that a variant will arise is very low indeed. In *Hydra*, the totipotent lineage may undergo an astronomical number of divisions before sexuality occurs. In dipterans such as *Drosophila*, it is appropriate to view the individual as a unique, genetically homogeneous, unit. It is highly unlikely that genetic variation will arise and gain access to the gametes within a single generation. In the hydroids, however, the number of divisions of the totipotent cell line intercalated between each sexual generation is so high that it is very likely that genetic variation will both arise and be inherited. In one case it is appropriate to view the individual as a genetically homogeneous unit and treat variation within a population without considering variation within individuals. In the other, it is not.

The individual as a genetically homogeneous unit is an ideal which is approximated to varying degrees in different taxa as a function of the mode of development of the particular taxa in question. It is of some importance, then, to identify phyletic patterns in developmental mode. This information is available. Embryologists of the late nineteenth century and, to a lesser extent, the twentieth century described patterns in development in great detail. Likewise, the embryologists of the twentieth century, with an experimental approach to development, have clearly pinpointed the timing and extent of multipotency in various lineages. This information is precisely that which the Modern Synthesis failed to accommodate.

21. The basal rate of change in base-pair sequence per cell is certainly not a constant, but likely varies as a complex taxon-specific function of the state of differentiation of the cell in question (e.g., B-lymphocytes in the mouse thymus have greatly accelerated mutation rates). Therefore, it is dangerous to presume that an estimate of genetic variation can be made solely on the basis of differences in the number of cell divisions, unless— as in this example—the differences are of astronomical proportions.

Taxon-specific differences in error rate may reflect the maintenance of traits which act either to generate or to minimize variation within the context of a particular developmental mode. In this regard, it is particularly intriguing that transposable elements (whose transposition generates considerable variation) are active only in the germ line of *Drosophila*, which makes few heritable divisions per generation, whereas transposable elements are active only in somatic cells of maize, which makes large numbers of heritable divisions per generation.

It is convenient to recognize three modes of development: somatic embryogenesis, epigenesis, and preformation.[22] In somatic embryogenesis, a distinct germ line is lacking. Rather, one cell lineage is capable of both participation in somatic function as a stem cell lineage and is also competent to give rise to gametes throughout ontogeny (e.g., plants). In contrast, organisms with epigenetic development possess a clearly differentiated germ line, but the germ line only appears after the primordia of major organ systems of the adult have become established (e.g., annelids). At the extreme end of this spectrum in the ontogenetic timing of the terminal determination of the germ line are organisms with preformistic development. Here the germ line is terminally differentiated in earliest ontogeny, often under direction of maternal-derived determinants deposited in the egg (e.g., nematodes).

The phyletic distribution of these modes of development is presented in Table 1.1 for all multicellular, cellular-differentiating taxa. The phyletic pattern is intriguing. Somatic embryogenesis is by far the most common mode of development. With the exception of the Volvacales, all multicellular representatives of the Kingdom Protista possess somatic embryogenic development. All members of the Kingdom Fungi and the Kingdom Plantae do as well. Only representatives of the Kingdom Animalia possess epigenetic and preformistic development, and these two modes of development are by no means ubiquitous. No fewer than nine animal phyla display somatic embryogenesis.

The ideal of the individual as an entity that may be treated as genetically uniform is at best an approximation. It is apparent that individuality is a derived character, approximated closely only in certain taxa. This fact is of substantial interest, for it means that not only is it inaccurate to consider the individual as the sole unit of inheritance in most taxa, but also that we have little assurance that it is appropriate to assume this to have been the case throughout geological time, even in those taxa in which individuality is now closely approximated.

---

22. The term "totipotent" is used here to refer to any cell which retains the capacity to produce both gametes and somatic cells.

TABLE 1.1   Phyletic Distribution of Developmental Mode[1]

Taxon[2]	Cellular Differentiation[3]	Developmental Mode[4]
PROTOCTISTA		
Phaeophyta	+/−	s
Rhodophyta	+/−	s
Chlorophyta	+/−	p
Ciliophora	+/−	s
Labyrinthulamycota	+/−	s
Acrasiomycota	+/−	s
Myxomycota	+/−	s
Oomycota	+	s
FUNGI		
Zygomycota	+	s
Ascomycota	+	s
Basidiomycota	+	s
Deuteromycota	+	s
PLANTAE		
Bryophyta	+	s
Lycopodophyta	+	s
Sphenophyta	+	s
Pteridophyta	+	s
Cycadophyta	+	s
Coniferophyta	+	s
Angiospermophyta	+	s
ANIMALIA		
Placozoa	+	s
Porifera	+	s
Cnidaria	+	s
Ctenophora	+	p
Mesozoa	+	p
Platyhelminthes	+	s, e, p
Nemertina	+	e
Gnathostomulida	+	u
Gastrotricha	+	p
Rotifera	+	p
Kinorhyncha	+	u
Acanthocephala	+	p
Entoprocta	+	s
Nematoda	+	p
Nematomorpha	+	u
Bryozoa	+	s
Phoronida	+	s
Brachiopoda	+	u

TABLE I.I (*cont.*)

Taxon[2]	Cellular Differentiation[3]	Developmental Mode[4]
ANIMALIA		
Mollusca	+	e, p
Priapulida	+	u
Sipuncula	+	u
Echiura	+	u
Annelida	+	s, e, p
Tardigrada	+	p
Onychophora	+	p
Arthropoda	+	e, p
Pogonophora	+	u
Echinodermata	+	e
Chaetognatha	+	p
Hemichordata	+	s, e
Chordata	+	e, p

1. After Buss (1983. *Proc. Nat. Acad. Sci. USA* 80:1387-1391).
2. Taxonomic divisions after L. Margulis and K. V. Schwartz (*Five Kingdoms*. San Francisco: Freeman, 1982). Although several of the phyletic distinctions advocated in this book are not traditional, this work has the merit of allowing the reader to consult a single text for an introduction to each group.
3. + = present in all known cases; +/− = presence in some cases and absence in others.
4. s = somatic embryogenesis; e = epigenetic; p = preformistic; u = unknown. The use of the terms epigenetic and preformistic refer only to the determination of the germ line and are not meant to recall the divisive debate on preformation of a century ago.

IV

It is one of the more remarkable oversights in biology that the patterns in developmental determinism just outlined failed to become incorporated into the Modern Synthesis. The lack of communication between geneticists, whose concern throughout this century has been with the mechanics of inheritance, and embryologists, whose concern has been with patterns in cellular differentiation, was clearly central. It is nevertheless astounding that the geneticists and natural historians forging the synthesis brought none of these developmental concerns to the fore.

At least two factors are of significance here. The first is largely a function of the organisms with which the framers of the Modern Synthesis worked. Virtually all of the early genetic discoveries were the work of botanists: the concept of mutation from deVries, the distinction between phenotype and genotype from Johannsen, linkage from Bateson's work on peas, multifactor inheritance from Nilsson-Ehle on wheat, and, of course, Mendel's peas. Later genetic work has also been profoundly influenced by botanists: for example, inversions and translocations from McClintock, genetic systems from Darlington, and polyploidy from Stebbins. However, sandwiched between these two periods of botanical pre-eminence was the central problem of the development and testing of the chromosome theory. This was the activity that dominated genetic work in the period just before the beginning of the Modern Synthesis (1910–1925). The central discoveries—that genes lie on chromosomes and that mutations occur in genes—were made using *Drosophila*. Accomplishments during this critical pre-synthesis period occurred in zoology. Similarly, the natural historians central in the development of the Modern Synthesis were largely zoologists. Dobzhansky worked on *Drosophila*; Mayr, Simpson, and Rensch on birds and mammals. Dipterans and vertebrates were the areas of expertise of the individuals who framed the synthesis. This zoological bias is particularly revealing. The geneticists who made the discoveries precipitating the synthesis, the mathematicians trained by these geneticists, and the natural historians recognizing their relevance to systematics and evolution, *all worked on organisms with preformistic development*. Fruit flies, mice, and humans are all organisms with early embryonic determination of the germ line—all are organisms in which the notion of the individual as a genetically homogeneous unit of selection is closely approximated. Thus neither geneticists nor evolutionists had reason to bring developmental concerns to the problems broached by the Modern Synthesis.

Just as the Modern Synthesis was unattended by embryologists, so did it pass untouched by botanists, mycologists, and zoologists of colonial invertebrates. The first botanical work applying the synthetic approach, G. Ledyard Stebbins's *Variation and Evolution in Plants*, did not appear until 1950.[23] Why did botanists, mycologists, and zoologists of

23. The expansion of the Modern Synthesis to clonal invertebrates is an even more recent development. That individual zooids or polyps within a colony are not equivalent to individuals in a genetic sense was only widely discussed following the publication in 1973 of *Animal Colonies* (Board-

colonial invertebrates remain silent during this period? Stebbins pinpoints the difficulty, noting that the:

> . . . problem is more serious in plants than in animals because in plants their germ plasm is not separate from somatic tissue, as Weismann showed for animals. New germ cells are differentiated every year from embryonic or meristematic cells. Until the molecular revolution, which demonstrated that DNA replication is independent of the environment, there was no theoretical reason for denying the inheritance of acquired characteristics.[24]

Plants, colonial invertebrates, and fungi all violate Weismann's doctrine, displaying enormous phenotypic plasticity and considerable ecophenotypic variation. These features worked against easy acceptance of the corollary of Weismann's doctrine—that acquired characteristics could not be inherited.[25] Weismann's doctrine of the continuity of the germ plasm predicted both that acquired modifications could not be inherited and that somatic tissues were not heritable. The latter prediction was borne out in organisms like fruit flies and rodents. However, this same issue of somatic inheritance was less than clear for many other organisms, leading to a considerable time lag in the acceptance of the former prediction.

V

The heritability of suborganismal variation in many organisms is a necessary consequence of known developmental pattern. It is not only fact, it is one of the capital facts of biology. A significant source of genetic variation in a broad spectrum of organisms is simply not incorporated into modern evolutionary theory. Yet the Modern Synthesis has en-

man, R. S., A. H. Cheetam, and W. A. Oliver, Jr., eds. Stroudsburg, Pa.: Dowden, Hutchinson, and Ross).

24. Stebbins, G. L. In E. Mayr and W. B. Provine, eds. *The Evolutionary Synthesis: Perspectives on the Unification of Biology*. Cambridge, Mass.: Harvard University Press, 1980:140.

25. G. Evelyn Hutchinson (pers. comm.), recalling Cambridge University in the early years of this century, attributes the rift between botanists and zoologists—whose pursuit of evolutionary synthesis has so often proceeded as if they were conceptually independent ventures—to this very issue.

joyed enormous heuristic success and from it have been wrought many generalizations of almost certain validity. The synthetic theory cannot be incorrect; it can only be incomplete.

The success of the Modern Synthesis, despite its inadvertent exclusion of developmental fact, is no paradox. The synthetic theory has succeeded because evolution has acted in all cellular-differentiating organisms in a manner such that individuality is approximated, albeit to varying degrees. Organisms appear as individual entities; evolution has yielded unambiguously discrete units. *The Modern Synthesis has stood as an enduring intellectual edifice for nearly a half-century because individuality has evolved.*

Why, though, has individuality evolved? Lacking in the Modern Synthesis, in evolutionary theory altogether, is an approach to this problem. Why, for example, does a mouse sequester its germ line and a hydroid not? Why are cell cycles synchronized in myxomycetes, but not in molluscs? Why do polychaetes have a variable number of segments and leeches a constant number? Why does nematode cell division cease in early ontogeny, whereas plant cells proliferate throughout ontogeny? Why do animals gastrulate, but follow different cleavage styles to arrive at this same endpoint? The Modern Synthesis has not generated a theory of ontogeny. Nor can it be expected to generate one: a theory which assumes individuality as a basal assumption cannot be expected to explain how individuality evolved. The synthesis Weismann sought remains elusive. Now, nearly a century after Weismann, it is self-evident that the missing chapters of the Modern Synthesis—the merging of genetics with development and the merging of development with evolution—remain the major tasks before us.

THE EVOLUTION OF DEVELOPMENT

Anything said on these questions lies in the realm of fantasy.
—L. H. HYMAN, 1959

Summary

Development has traditionally been viewed as a cooperative enterprise. The zygote gives rise to clones of cells which, by a complex interplay of mutual interactions, faultlessly orchestrate the production of a new individual. While the cooperative model is an appealing characterization of development in extant taxa, it is inappropriate to view the processes which gave rise to developmental programs as cooperative. The opposite is more likely the case. At the dawn of metazoan life, the germ line was not determined at the outset of ontogeny. Those cells which ultimately gave rise to new individuals had to earn their position in the germ line by surviving a somatic selective filter in the form of competitive interactions with other cell lineages comprising the same individual. Those variant lineages which successfully outcompeted their neighbors could have had one of two broad classes of effects on the individual. The variant may have disadvantaged the individual harboring it and disappeared or, just as fortuitously, advantaged the individual harboring it and flourished. Subsequent variants may have had a net positive, neutral, or negative effect on the rate at which further variants arose and hence acted differentially to preserve or obscure the effects of those variants preceding them. The thesis developed here is that the complex interdependent processes which we refer to as development are reflections of ancient interactions between cell lineages in their quest for increased replication. Those variants which had a synergistic effect and those variants which acted to limit subsequent conflicts are seen today as patterns in metazoan cleavage, gastrulation, mosaicism, and epigenesis.

The conservatism of early metazoan ontogeny is held here to reflect the interplay between selection at the level of the individual and selection at the level of the cell lineage, under the strong influence of an ancestral constraint. The protist groups which gave rise to metazoans were taxa bearing only a single, uncommitted, microtubule organizing center per cell. As microtubule organizing centers are required for both cell division and ciliation, the possession of only one such structure precludes simultaneous cell division and ciliation. This ancestral constraint limited the capacity of a free-living embryo to simultaneously swim by ciliary motion and to continue develop-

ment via cell division. Any cell lineage which abandoned ciliation and moved into the ciliated blastula increased its own rate of repli-cation by doing so. In the process, the variant discovered gastrula-tion and invented germ layers. Only certain cells undergo this trans-formation in extant organisms, because cytoplasmic determinants deposited in the egg by the mother act to limit the further expression of variants which abandon the somatic function of locomotion for continued cell division. Limits, however, are placed on the extent of maternal control, as an egg can hold only a limited range of cyto-plasmic instructions and yet dole them out equally to each cleavage product. Unequal cleavage arose, but only in those taxa in which the free-living embryo is released after gastrulation. Selection at the level of the cell lineage for increased replication—in the form of cil-iated cells of the free-swimming blastula abandoning ciliation for enhanced cell division—established gastrulation and subsequent limits on the rate of further variation—in the form of maternal cy-toplasmic determinants—established unequal cleavage and variant styles of gastrulation. A perspective of selection operating on both units of selection in synergism and conflict provides a simple predic-tive evolutionary model for phylogenetic distribution of the various styles of cleavage, gastrulation, and maternal predestination dis-played by metazoan embryos.

The diversity of late ontogeny is similarly addressed. Following a variable period of maternal control of embryonic cell fate, the met-azoan embryo becomes organized into one of several discrete bauplans via interactions between embryonic cell lineages. The principal epi-genetic interactions defining cell fate—those of induction, compe-tence, and cell death—are all interactions in which one cell lineage acts to limit the replication of another, while enhancing its own. The fact that embryos develop by epigenesis is prima facie evidence that these very programs represent interactions between variant cell lineages arising in the course of ontogeny of ancestral forms. Selection at the level of the cell lineage also favored variant programs whose phenotypic effect was to limit further variation. Among these were cell lineages which enforced terminal differentiation of the germ line. Somatic lineages continued to proliferate, while the mitotic activity of the germ line became severely limited. Individuality evolved as germ line sequestration acted to restrict the origin of new heritable variation in the course of ontogeny. Further evolutionary innnova-tion became limited to phylogenetic shifts in the relative timing of germ line determination (i.e., heterochrony), which differentially permits or precludes heritable intrusions in the course of ontogeny. Looking at the evolution of development as the interplay between se-

lection on the level of the cell lineage and selection on the level of the individual not only predicts that metazoans should develop by epigenesis and that their genomes need code only for relative competitive abilities, but it also links these traditionally reductionistic explanations to observations of the rapid appearance of metazoan bauplans, the subsequent lack of elaboration of new bauplans, and the eventual modification of existing bauplans via heterochrony.

While Weismann recognized that terminal differentiation precludes cells from participation in the production of a new individual, his inheritance theories blinded him to the fact that any dividing cell, independent of its state of differentiation, retains the capacity to generate and transmit variation to its lineal descendants. As with any self-replicating system, variants which serve to enhance the replication rate of the cell lineage will be favored. The selection of variant lineages with enhanced replication rates proceeds independent of whether the variant favors the individual harboring it. In adopting Weismann's doctrine, the synthetic theory of evolution explicitly excluded from consideration the origin of variants in the course of ontogeny and selection of those variants within the somatic environment. In so doing, the synthetic theory proved capable of incorporating patterns of organismal adaptation, but sacrificed the ability to link those explanations in a simple causal scheme to the underlying developmental processes giving rise to them. The expansion of the synthetic theory to include consideration of multiple units of selection is sufficient to link the patterns traditionally the province of holistic biologists to their underlying reductionist explanations via a thoroughly straightforward Neo-Darwinian cognitive style.

THE CONSERVATISM OF EARLY ONTOGENY

I

The ontogeny of metazoans begins in a highly stereotyped fashion. Whether the resulting individual is to develop into a coral or a clam, the initial stages of metazoan development display a marked similarity. The zygote typically cleaves in either a spiral or radial fashion to yield a spherical ball of cells, the blastula, which undergoes a series of morphogenetic movements producing a multilayered structure, the gastrula. The zygote of a coral cleaves somewhat differently from that of the clam, and the gastrula is formed by rather different patterns of cell movement, but both undergo manifestly similar transformations of cell division, differentiation, and movement. From these remarkably uniform early ontogenetic events, the foundations are laid for radically different adult structures. Nowhere in the early ontogeny of the coral or the clam are their differences as adults apparent; the essential processes of cleavage, blastulation, and gastrulation mask the diversity of the adult structures into which the embryos ultimately develop.

As von Baer first recognized and codified in his phylogenetic laws, the uniformity of early ontogentic events within and between taxa reflects the inherent difficulty of modifying an existing pattern once it is established, coupled with common ancestry. It is axiomatic that a random alteration introduced early in ontogeny will likely be manifested in a cascade of subsequent morphogenetic events, whereas a modification introduced later in ontogeny can have relatively minor effects.[1] The validity of this interpretation can hardly be doubted. A random error in the manufacture of the central processing unit of my computer would unquestionably preclude any hope of my using it to write, while the various marketing decisions reflected in the design of the exterior case have provided me with only minor inconveniences.

1. Experimental support is found, for example, in the effect of cells homozygous for zygotic lethals expressed at different stages in ontogeny (Ripoll, P. and A. Garcia-Bellido. 1979. *Genetics* 91:443–453) and in ontogenetic effects following a standard x-ray dose (Wilson, J. G. and F. C. Fraser. *Handbook of Teratology*. New York: Plenum Press, 1977).

The difficulty of tinkering with a complex existing pattern has traditionally served as a sufficient explanation for the extreme conservatism of early ontogenetic events. The traditional explanation is so clearly correct, so comfortable a notion, that it is quite easy to overlook the fact that it does not provide any explanation whatsoever for the patterns themselves. We simply do not know why animals undergo the specific patterns of early development that they do—why metazoans cleave as they do, why blastula are formed, why gastrulation occurs, or why different taxa display differing combinations of these early morphogenetic events.

II

Evolution can only yield variants of that which it has already produced. Previous design constrains the range of subsequent variation. As François Jacob succinctly puts it: "Not just any organism is possible."[2] Constraints governing metazoan ontogeny must be found in their historical antecedents, the protists. Protist cell architecture and function are exceedingly diverse and often very specialized. The intricate fibrillar cell membrane of ciliates, the exquisite skeletons of radiolarians, and the peculiar synchronized nuclear division in myxomycete plasmodium come to mind as examples. These, and innumerable other modifications, provided the opportunity to exploit particular environmental conditions, but no doubt also limited the potential of the taxa for subsequent evolutionary modification.

A single-celled protist must simultaneously express specialized modes of locomotion, feeding, and behavior, and yet retain the capacity for indefinite cell division. Metazoans, however, have no similar requirement. The opposite seems more commonly to be the case. Highly differentiated metazoan cells are often characterized by a limited potential for continued differentiation, or even for continued cell division. Neuroblasts, for instance, can give rise only to other neuroblasts or neurons, and neurons cannot divide at all. Cellular specialization in metazoans is often incompatible with further replication. If the specialization of protist cell architecture represents modifications imposed by conflicting

2. Jacob, F. *The Possible and the Actual*. New York: Pantheon Books, 1982:21.

demands of subcellular elaboration and continued replication, then metazoans inherited these same constraints. Metazoans, however, were free to resolve such conflicts by an entirely different means, by doling out cellular specialization to differentiated cell lineages while retaining unspecialized cells for further differentiation.

The incompatibility of increasing cellular complexity with maintenance of intracellular cooperation and continued division is by no means a fresh idea. Yet, this time-honored generalization is furthered only by an attempt to define those cellular transformations that are possible and using knowledge of these constraints to interpret that which has actually evolved. Cell biologists, indeed the histologists of the late nineteenth and early twentieth centuries,[3] have long been aware that the mitotic "dance of the chromosomes" bears a curious and fascinating relationship to the development of cilia and flagella. The mitotic spindles of metazoans and undulipodia (i.e., cilia or flagella) of protists are undoubtedly homologous. Both undulipodia and the mitotic spindle are long hollow structures of characteristic size, composed of a series of closely related proteins called tubulin. All such structures grow from "microtubule organizing centers,"[4] variously called basal bodies, centrioles, nucleus-associated organelles, or kinetosomes,[5] depending upon their position in the cell and the specific structures they create. Since undulipodia and the mitotic spindle both require microtubule organizing centers, the capacity for simultaneously accomplishing the functions of both organelles is a critical concern. As Lynn Margulis has recently emphasized, at issue is the simultaneous need for an organism to move through a fluid medium using cilia or flagella and to divide using a mitotic

3. Wilson, E. B. *The Cell in Development and Heredity.* New York: Macmillan, 1925.
4. Pickett-Heaps, J. D. 1969. *Cytobios* 1:257–280.
5. Few organelles have been given so many different names as those from which microtubules arise (e.g., centrum, centriole, centrosphere, attraction sphere, attractophores, idiozome, blepharophast, kinetosome, spindle pole body, nucleus associated body, etc.). Early in this century, nearly every cytologist used a different term for the centriole and, for a time, debate over terminology replaced debate over substance (Boveri, T. 1895. *Verhandl. physik.-med. Ges. Wurzburg* 29:1–75). Today, a similar diversity exists in the terms for the structures occurring at the poles of mitotic spindles. For simplicity, I chose to refer to all such structures using Pickett-Heaps's term (*op. cit.*), the microtubule organizing center.

a

b

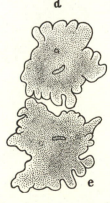

c

d

e

spindle. Unless a cell possesses microtubule organizing centers capable of performing both tasks, or possesses multiple microtubule organizing centers per cell, the cell's functional range will be constrained.[6]

In some protistan taxa, including the Sacrodinia, Myxosporida, Rhodophyta, and acrasid slime molds, only a single microtubule organizing center is present per cell. These groups can divide mitotically, but undulipodia are absent.[7] Cell movement is either lacking entirely or occurs via mechanisms other than ciliation (Figure. 2.1). Motility in taxa with a single microtubule organizing center is not necessarily precluded, though. If the microtubule organizing center is not terminally committed to a single task, then both functions can occur—albeit not synchronously. For example, the haploid phase of the myxomycete life cycle involves proliferation of amoeboid cells. These cells, however, retain the capacity to develop motile flagellated cells. If a myxomycete clone growing on solid substrata is plunged into water, the amoeboid cells will differentiate flagella and begin to swim. When solid substratum is reached, the flagella disappear, cells assume an amoeboid form, and division follows. The flagellated cell, however, does not divide while flagellated.[8] In taxa with a single microtubule organizing center, simultaneous ciliation and cell division appear to be mutually exclusive alternatives.

Movement and cell division can occur simultaneously in a variety of protist taxa. In certain hypermastigotes and opalinids, extranuclear microtubule organizing centers act as organizers of both the spindle and the flagella. In the flagellate *Barbulanympha*, for example, the microtubule organizing

6. Lynn Margulis (*Symbiosis and Cell Evolution*. San Francisco: Freeman, 1981) raises this point in the context of a phyletic overview of protist cell division. While much of the discussion of protist cell division detailed above is derived from (and/or inspired by) her review, the conclusions reached are independent of her fascinating suggestion that tubulin-based intracellular structures are evolutionary relicts of past symbiosis between primitive eukaryotes and spirochetes.

7. These taxa may have lacked cilia in their primitive state or may have lost cilia secondarily. Margulis (*op. cit.*), who favors a spirochete origin for intracellular tubulin-based structures, has argued in detail for the latter interpretation.

8. Flagellated myxomycete cells can, in fact, divide. However, to do so they must first withdraw their flagella and redifferentiate the structure following mitosis (Schuster, F. 1965. *Prostistologica* 1:49–62; Olive, L. S. *The Mycetozoans*. New York: Academic Press, 1975).

FIGURE 2.1

Stages (a–e) in cell division of *Amoeba proteus*. (From Liesche, centers are long filiform structures with one pole anchored at the base of the flagella and the other pole functioning as an extranuclear division center (Figure 2.2). In mitosis, one pole of the microtubule organizing center, bearing centrosomes, gives rise to astral rays which meet to form a central spindle, while the other pole gives rise to a new microtubule organizing center. Shortly after this new center is formed, it gives rise to a complete set of new flagella, these being fully developed by early anaphase. Here, cell division and motility are accomplished simultaneously, as the microtubule organizing center can simultaneously produce the spindle and the organelles of locomotion.[9]

Other protist groups solved the need for simultaneous movement and cell division in a rather different way. The

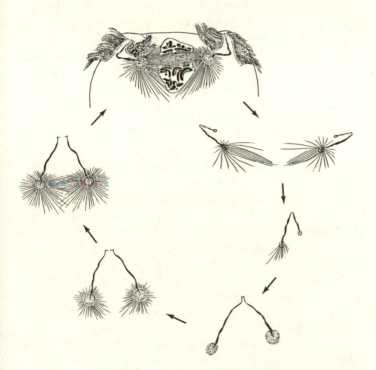

FIGURE 2.2   Life cycle of the centriole of *Barbulanympha*. (After Cleveland, 1938, 1971.)

9. Cleveland, L. R. In L. Levin, ed. *The Cell in Mitosis*. New York: Academic Press, 1963:3–31.

ciliates, for example, are a large group of 8,000 or more species. Cells may be quite large and are often uniformly ciliated. An autonomously replicating microtubule organizing center occurs at the base of each cilium. Microtubule organizing centers, however, are absent in the cell nucleus. No ciliate macronucleus undergoes mitosis. Rather, macronucleus division is a complex and inefficient process involving a distinctive behavior of a large number of copies of short pieces of DNA.[10] Ciliates can divide and move at the same time, but division of the macronucleus is a seemingly disordered process. While effectively solving the difficulty of simultaneous movement and cell division, the ciliate macronucleus lacks a spindle and must divide amitotically.

Yet another alternative to the demand for simultaneous cell division and locomotion is for the cell to divide by mitotic mechanisms, but to maintain an additional microtubule organizing center committed to the development of undulipodia. The Euglenophyta, Cryptophyta, and Chlorophyta frequently possess multiple microtubule organizing centers and are capable of accomplishing simultaneous cell movement and mitotic cell division by using some microtubule organizing centers exclusively as organizers for undulipodia, whereas others are differentiated as intranuclear division centers. In the trypanosome *Trypanosoma brucei*, for example, new cytoplasmic microtubule organizing centers appear just before cell division and elaborate undulipodia (Figure 2.3). Intranuclear division centers then direct mitosis. Here seemingly autonomous microtubule organizing centers carry out their distinct tasks. Simultaneous cell division and locomotion require only that the replication of cytoplasmic microtubule organizing centers be synchronized with nuclear division.

Some groups, however, apparently never resolved the problem of simultaneous cell division and ciliation. Undulipodia disappear when the mitotic spindle forms. This process is particularly dramatic in heliozoans, a group characterized by large tubulin-based axopods. These locomotory and feeding structures are withdrawn prior to cell division; the microtubules disappear to below the resolution of the electron microscope. Shortly thereafter, the now immobile cell begins the division process, with the microtubules re-

10. Gabriel, M. 1960. *Am. Nat.* 94:257–269.

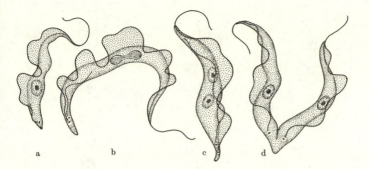

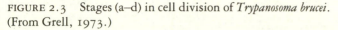

FIGURE 2.3 Stages (a–d) in cell division of *Trypanosoma brucei*. (From Grell, 1973.)

polymerizing immediately after cytokinesis. A mitotic spindle is formed, either using distinct plaques or the former base of the axopods as microtubule organizing centers. The cell divides, and the axopods are redifferentiated in the daughter cells (Figure 2.4).[11] Here again cell division and motility seem mutually exclusive alternatives, separated in time. The resorption of axopods may represent either a central role for cytoplasmic microtubule organizing centers in spindle formation or, more simply, the need to localize microtubule organizing centers at the site of nuclear division to ensure an even distribution of cytoplasmic microtubule organizing centers to daughter nuclei. In either case, the result is the same: simultaneous cell division and locomotion is precluded.

In certain protist groups, cell division and locomotion can occur synchronously; in others they cannot. The operation of this constraint in the axopod-bearing heliozoans is particularly revealing. Heliozoa are believed to have given rise to the choanoflagellates, a group of solitary or colonial flagellates which bear a distinct collar surrounding the base of a single flagellum.[12] The choanoflagellates are traditionally regarded as a link between the protozoa and the sponges, as sponges bear collared cells (the choanocytes) that are remarkably similar in both structure and function to choanoflagel-

11. Cleveland, L. R. 1935. *Science* 81:597–600; Cleveland, L. R. 1938. *Biol. Bull.* 74:1–24.
12. Hyman, L. H. (*The Invertebrates: Protozoa through Ctenophora*. New York: McGraw-Hill, 1940) argues that the collar of choanoflagellates arose via the fusion of axopods.

FIGURE 2.4

Cell division in the heliozoan *Actinophrys sol*. Note that the nucleus (n) comes to occupy a central position in the cell and the mitotic spindle (s) begins to form only after the axopods (ax) are withdrawn. (After Belar, 1922.)

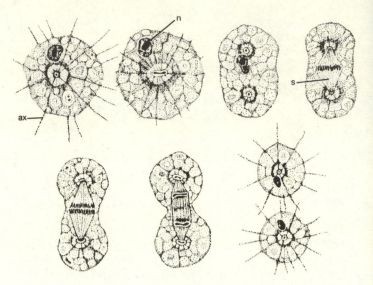

lates.[13] In fact, one colonial choanoflagellate, *Proterospongia*, even forms a gelatinous coat with collared cells on the outer surface and dividing amoeboid cells in the interior, recalling the typical sponge structure of mesenchyme, choanocytes, and amoebocytes (Figure 2.5).[14]

While many protist taxa overcame the ciliation constraint, *those protists giving rise to metazoans did not*. Metazoans

13. In addition to their prominent role in sponges, cells with distinct collars occur in the larval stages of several metazoans. The collared cells of bracheolaria larva of echinoderms and those in the planulae larva of corals have been interpreted as phylogenetic relics (Norrevand, A. and K. G. Wingstrand. 1970. *Acta Zool., Stockh.* 51:249–270; Lyons, K. M. 1974. *Nature* 245:50–51). These cells, however, bear a number of ultrastructural differences, particularly in the flagellar apparatus, which distinguish them from the collared cells of both sponges and choanoflagellates (Hibberd, D. J. *op. cit.*).

14. *Proterospongia* certainly does not represent a literal stage in the phylogeny of sponges. Nevertheless, *Proterospongia* illustrates an important point regarding the phylogeny of sponges that has not, to my knowledge, been clearly articulated. The constraint against simultaneous cell division and locomotion in choanoflagellates requires that a choanoflagellate lose its collar and flagellum to divide. The fact that some colonial choanoflagellates resolved this problem by producing amoeboid dividing cells provides a clear mechanism to generate the typical sponge association of nondividing, collared, flagellated choanocytes and dividing amoebocytes. In this view, the otherwise peculiar sponge pattern of choanocyte differentiation to amoebocytes and vice versa is a natural consequence of a primitive constraint on cell division.

inherited the constraint limiting simultaneous mitosis and
ciliation. As Pitelka puts it: "Those organisms that ultimately were to spawn the metazoa retained the kinetosome/
centriole [microtubule organizing centers] as a versatile, po-
larized, morphogenetic center but did not enslave it, and
their genetic systems, in the service of immediate morphologic elaboration."[15] Metazoan mitosis proceeds via extranuclear organizing centers (Figure 2.6). Undulipodia are re-
sorbed prior to the initiation of mitosis. Microtubules used
to form the spindle are polymerized from previously synthesized tubulin, and ciliation reappears only after the spindle
has disappeared.[16] No metazoan cell bearing undulipodia or
other microtubule-based structures divides while ciliated.[17]
The diversity of these structures is enormous. Besides the

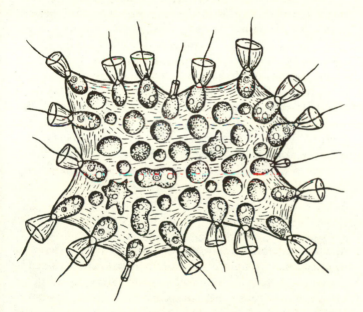

FIGURE 2.5

Proterospongia haeckeli, a
choanoflagellate with
nondividing flagellated
cells occupying the ex-
terior of the colony and
dividing amoeboid
cells occupying the in-
terior. (From Saville
Kent, 1882.)

15. Pitelka, D. R. *Electron-Microscopic Structure of Protozoa*. New York:
Macmillan, 1963:231.
16. Certain mitotic movements are attributable to polymerization of tubulin, presumably derived in part from cilia, into the elongating microtubules of the spindle (Mazia, D. In K. B. Warren, ed. *Formation and Fate
of Cell Organelles*. New York: Academic Press, 1967:39–54; McIntosh,
R., P. K. Helpler, and D. Van Wie. 1969. *Nature* 224:659–661).
17. This remarkable pattern appears to have been first recognized by Lynn
Margulis (*op. cit.*).

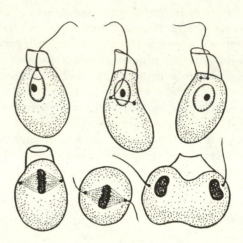

FIGURE 2.6   Cell division of the choanocyte of the calcareous sponge *Clathrina*. Note the lack of simultaneous cell division and flagellation. (From Beklemishev, 1969.)

undulipodia of epithelial cells, tubulin-based structures include the axons and dendrites of nerve cells, the kinocilia of invertebrate statocysts and vertebrate ears, the many mechanoreceptors of insects and crustaceans, and the tails of sperm.[18] No ciliated metazoan cell, no metazoan nerve cell, no metazoan mechanoreceptor, and no metazoan sperm ever divides. A protistan legacy apparently remains: ciliated metazoan cells do not divide when ciliated.

III

Metazoan life histories necessarily cycle through a period of multicellular proliferation and differentiation punctuated by periodic generative single-cell stages.[19] Since the adult organism will always experience some finite threat of mortality, continued existence of the organism requires that it periodically produce and disperse propagules to reestablish the life cycle at other locations. The propagules of ancient metazoans cannot have been exceptions—their embryos must have moved through a fluid medium. Movement of a propagule may be entirely passive or the propagule may possess

18. Dustin, P. *Microtubules*. Heidelberg and New York: Springer-Verlag, 1978.

mechanisms for small-scale controls over its own movement. The latter requires that the embryo be ciliated.

The gametes of the many extant representatives of the simplest metazoan phyla are shed directly into the ocean. Here fertilization takes place and the embryo undergoes a rapid series of cleavage divisions giving rise to the morula, a solid ball of undifferentiated cells. The morula becomes ciliated and assumes the typical spherical shape of the blastula. Herein lies a difficulty. The protistan legaacy precluding simultaneous cell division and ciliation demands that the uniformly ciliated embryo can undergo no further cell division. The blastula can hardly continue to develop while moving, for its surface is covered with ciliated cells, each of which is unable to divide. In this respect the ciliated embryo has inherited a severe liability from its protist ancestors. If the embryo is not to abandon dispersal at this early stage of ontogeny, mechanisms must have evolved which allow it to escape its protistan past. A mechanism must exist by which the embryo may simultaneously continue to move, yet continue to develop.[20] How then can an embryo, covered with cells incapable of dividing, continue to develop while moving?

Some metazoans never did resolve this conflict; rather, they abandon dispersal at this stage. For example, the morula of the tectractinomorph sponge *Tethya serica* never develops cilia, bypassing both the blastula and gastrula stages.[21] The morula simply drops to the surface and develops directly into an adult sponge. A similar situation is found in the stalked scyphozoans *Thaumatoscyphus distinctus*[22] and *Haliclystus octoradius*.[23] Gametes are released into the sea, fertilization occurs, and cleavage is initiated. The embryo becomes a morula, but instead of going on to form a blastula in the typical jellyfish fashion, the embryo never becomes ciliated. This modified larva drops to the sea floor, moving via contractions and expansions of the body wall, and metamorphoses without a ciliated, free-living larval

19. Bonner, J. T. *Size and Cycle*. Princeton University Press, 1965.
20. To the extent that the ancestral metazoan stock was a flagellated pelagic organism, this dilemma was first faced not by the embryos of ancient metazoans, but by their phyletic precursors.
21. Watanabe, V. 1957. *Natur. Sci. Rep. Ochanomizu Univ.* 8:97–104; Watanabe, V. 1960. *Bull. Mar. Biol. Stat. Asamushi Tohoku Univ.* 10:145–148.
22. Hanaoka, K. 1934. *Proc. Imp. Acad.* 10:117–120.
23. Wietrzykowski, W. 1912. *Arch. Zool. Exp. et Gén., Sér.* 5, 10:1–95.

stage.[24] An improvement on this scheme is seen in two other sponges, *Polymastia robusta*[25] and *Raspailia pumila*.[26] Here the morula does develop into a blastula, which sinks to the seafloor, flattens slightly, and adopts a crawling motion. Without further cell division, the larva searches until an appropriate location is found and metamorphosis into an adult sponge begins. These organisms provide no solution to the problem. Rather they illustrate the failure of some taxa to have found a way to simultaneously develop and disperse.[27]

Most metazoans, however, have solved the dilemma. The development of a morula into a ciliated blastula involves the retention (or production) of a small number of cells which are not ciliated (Figure 2.7). These cells, however, can proliferate in only one direction. Since the primitive form is a sphere traveling in water, the proliferating cells cannot grow into the water column without producing a mass of cells which would certainly increase drag on the embryo, interfering with further movement. Neither can the proliferating cells grow over the surface of the sphere, for to do so would obliterate the ciliated surface necessary for locomotion. Only one alternative remains. *The only remaining place for these cells to move is into the sphere itself.* The movement and subsequent proliferation of cells from the blastular surface into the center of the blastular sphere is gastrulation. Animal gastrulation is the metazoan solution to the requirement of simultaneous development and movement.

Gastrulation, in its most simple form, is a necessary consequence of the protistan constraint against simultaneous ciliation and cell division, coupled with the need for development to continue while the embryo is dispersing. The problem is as severe today as it was eons ago. Many sponges, cnidarians, echinoderms, and nemertines still produce free-living ciliated blastula and require gastrulation to develop while moving. Gastrulation in its simplest form occurs by

24. The absence of blastula and/or gastrula stages in these forms should not be mistaken as primitive traits.
25. Borojevic, R. 1969. *Cah. Biol. Mar.* 8:1–6.
26. Levi, C. 1956. *Arch. Zool. Exp. Gen.* 93:1–181.
27. The observation that both cnidarians and sponges display parallel retrograde larvae is particularly interesting, in light of the lack of obvious morphological traits linking the two phyla in simple phylogenetic progression. This observation strongly suggests that the ciliation constraint has been sufficiently powerful as to guide selection along similar lines in both taxa.

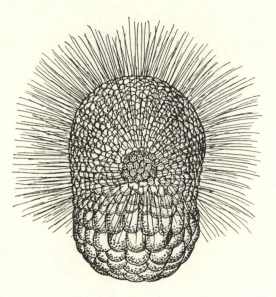

FIGURE 2.7   Amphiblastula of the sponge *Leucosolenia*. (From Minchin, 1896.)

unipolar invagination (emboly), whereby cells at the vegetal pole migrate into the interior (Figure 2.8). The surface of the blastula acquires an identation in the zone of proliferation, just as if a finger had been pushed into a balloon.

While emboly is common, the process has been modified in various ways. In the sea anenome *Halcampa duodecimcirrata*, gastrulation begins as a typical invagination, followed by the detachment of some blastoderm cells from this layer into the blastocoele. Here the initial phase of invagination is transformed into introgression.[28] The stage of emboly is lost in many groups. For example, in the hydroid *Aequorea forskalea*, ciliated blastoderm cells at the vegetal pole of the blastula wander into the central cavity of the blastula, without emboly, and proliferate to form a secondary cell layer (Figure 2.9). In other forms, for example the sponge *Leucosolenia blanca*, introgression is not localized at one pole. Here, a small number of initially ciliated cells at various points over the blastular surface lose their cilia, become amoeboid, and begin to divide, establishing a two-layered organism by multipolar introgression (Figure 2.10).

28. Nyholm, K. G. 1949. *Zool. Bidr. Uppsala.* 27:465–506.

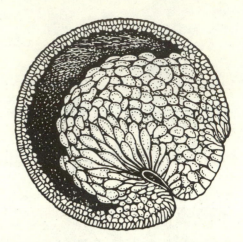

FIGURE 2.8   Gastrulation by emboly in an amphibian embryo. (From Holtfreter, 1943.)

Whether gastrulation occurs by invagination, or by unipolar or multipolar introgression, the result is the same: a single-layered ciliated embryo incapable of further development is transformed into a two-layered organism capable of further movement via the ciliated ectoderm and of further development by the unciliated endoderm.

While emboly, unipolar introgression, and multipolar introgression are frequent in forms which retain a ciliated blastula, they represent only a fraction of the existing styles of gastrulation. Many taxa have abandoned the habit of shed-

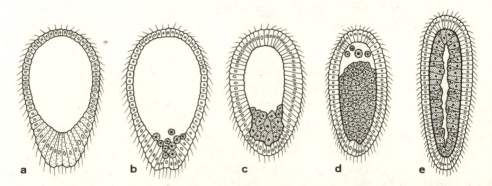

FIGURE 2.9   Stages (a–e) of gastrulation by unipolar introgression in the hydroid *Aequorea forskalea*. (From Claus, 1883.)

ding their gametes directly into the sea. Fertilization is internal and the embryo is brooded for varying lengths of time.
The simplest modifications are seen in taxa in which the embryo, while retained in the mother during earliest ontogeny,
is eventually released as a blastula. Such forms necessarily
retain their ciliated surface epithelium. In contrast to the
free-living morula, however, they are not faced with the immediate need for movement, and they develop somewhat
differently. Here the morula gives rise to a blastula which is
initially without cilia. The blastoderm cells undergo a series
of mitotic divisions with radial spindle orientations, as in
the hydroid *Clava squamata* (Figure 2.11). The blastula acquires cilia and is released from the parent. The blastula retains from the outset several unciliated cells in the interior,
often producing a solid or stereoblastula. This structure may
develop further, either through the delamination of the interior cells into a definitive tissue layer, or by the combina-
tion of this process with invagination or introgression.
Again a two-layered structure is produced, again simultaneous ciliation and cell division is precluded, but again the
larva can both swim and continue to develop.[29]

Even greater variations are known. In a large number of
metazoan phyla, the embryo develops within maternal tissues (in thin-walled, free-living egg capsules), or attached to
the sea floor (in thick-walled, resistant shells or cocoons),
until gastrulation has occurred. The embryo only becomes
ciliated after a proliferating internal layer has become established and the gastrula, or an even more highly developed
larva, hatches. In this case, the requirement of a uniformly
ciliated blastula is removed entirely and gastrulation may
take place by quite different means. Whereas the only cells

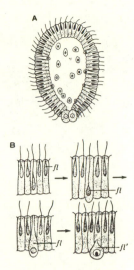

FIGURE 2.10

Gastrulation by multi-
polar introgression (A)
in the sponge *Leucoso-
lenia blanca*. Note that
the cells undergoing
introgression (B) lose
their flagellae prior to
movement into the
blastocoele. (From
Minchin, 1900.)

29. Haeckel saw the coeloblastula as the primitive metazoan, but this
view was overturned by Metschnikoff. Most modern phylogeneticists
(e.g., Hyman) follow Metschnikoff in seeing the stereoblastula, a common stage in both sponges and cnidarians, as primitive. Metschnikoff
based his argument of the ancestral status of the stereoblastula on the fact
that invagination is rare in hydroids. However, most hydroids and several
sponges retain the embryo within maternal tissues prior to release. In the
argument presented here, this brief interval of maternal brooding allows
the blastula a temporary respite from the ciliation constraint, hence permitting the *secondary* development of a stereoblastula. A consideration of
the ciliation constraint, therefore, removes the principal opposition to
Haeckel's blastula. Whether this vindicates Haeckel's phylogenetic infer-
ences of early metazoan ontogeny is, of course, another matter altogether.

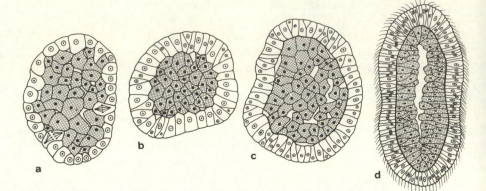

FIGURE 2.11   Stages in the development of the stereogastrula (a–d) in the hydroid *Clava squamata* (From Harm, 1903.)

of a ciliated, free-living blastula that can proliferate are those which gain access to the interior, development within an egg capsule places no such constraint on cell proliferation. Cells on the surface of one pole of the unciliated blastula can divide at a faster rate than those at the opposite pole, eventually enclosing the former to produce a two-layered organism (Figure 2.12). This mode of gastrulation, known as epiboly, is clearly precluded in the free-living blastula, for it would result in the covering of the very cells necessary for locomotion.[30] Taxa with encapsulated eggs from which free-living post-gastrula larvae hatch include the flatworms, molluscs, annelids, arthropods, and a variety of minor metazoans. The removal of the requirement for a free-living blastula has allowed the evolution of modes of gastrulation prohibited in ancestral groups.

Several taxa have entirely lost the need for an embryo which moves freely through seawater; they have adopted instead some form of direct development. In some forms which brood their embryos to a crawling stage, where the egg is

30. Epiboly is usually interpreted as a modification of gastrulation required by yolky eggs. Early cleavage products are simply too large to migrate effectively. Hence, small cleavage products enclose them. The influence of yolk content on modes of gastrulation is not questioned here. Rather, these arguments address the causes underlying the evolution of high yolk content and the forces restraining the evolution of high yolk reserves in groups with a free-living blastula stage.

composed of large stores of yolk, all hints of gastrulation are lacking. For example, in the rhabdocoele and triclad flatworms, early development proceeds in a most peculiar fashion (Figure 2.13). Each shell contains a zygote and several yolk cells; cleavage of the zygote leads to blastomeres which separate and adopt amoeboid motion. These amoeboid cells move about the yolk cells ingesting their contents. Only later do these cells and their daughters reaggregate to begin construction of the worm itself. No hint of a blastula or a gastrula occurs. Despite this unusual early ontogeny, few people would argue that the rhabdocoeles and triclads are anything but flatworms derived from other turbellarian flatworm orders which produce quite normal blastula and undergo thoroughly unremarkable patterns of gastrulation.

While von Baer's realization that early transformations in ontogeny were refractory to evolutionary modification is valid, his explanation alone is inadequate. Early ontogenetic events in some taxa have indeed undergone substantial evolutionary modification, as the divergent modes of development in some flatworms clearly illustrate.[31] Early ontogenetic events have been modified, but only in those taxa in which modification was *possible*. An ancestral constraint and an active selection pressure has restrained, and continues to restrain, the modification of early ontogeny. In free-swimming embryos, where the constraint is relevant and the selective pressure operative, evolution has produced only mi-

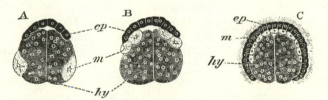

FIGURE 2.12   Stages of gastrulation (A–C) by epiboly in the flatworm *Leptoplana tremellaris*. The large macromeres (hy) and mesoblasts (m) come to be surrounded by the smaller, more rapidly dividing epiblasts (ep). Note that the epithelium does not become ciliated until gastrulation is complete. (From Balfour, 1881.)

31. Modifications of early ontogeny are by no means limited exclusively to parasitic flatworms; rather they occur in several taxa which lack a free-swimming stage (e.g., insects and nematodes).

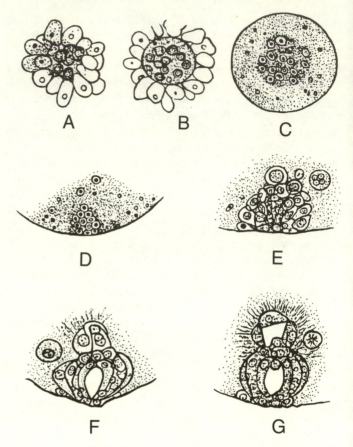

FIGURE 2.13 Early development of the triclad flatworm *Dendro-coelum lacteum*. In the early cleavage stages (A) the fertilized ova migrates among yolk cells, absorbing their contents (B). After forming an external membrane (B–C), the free-living blastomeres aggregate at the ventral side of the embryo (D) and begin to elaborate the embryonic pharynx (e–g). Finally, the embryo absorbs the remaining yolk and proceeds to elaborate the various tissue layers of the adult. No hint of any structure akin to a blastula or gastrula is found in early development. (From Iijima, 1884 and Mattiesen, 1904.)

nor variations on the theme of emboly. In embryos developing within maternal tissues or within encapsulated eggs, where the selective pressure is temporarily relaxed, evolution has produced variant styles precluded in instances where selection is continuously operative. Finally, in direct developing embryos, where the selective pressure is permanently removed, all hints of ancestral ontogeny are lost.

IV

The dividing zygote is a clone. Just like a clone of unicellular organisms, embryonic cell divisions pass on any genetic variation which has arisen in previous divisions. Likewise, any genetic variant which gives rise to daughter cells with an increased replication rate will be favored, whether the variant occurs within a developing embryo or in a clone of unicellular organisms. Although the vast majority of variants which arise will prove to be detrimental, a small percentage will prove to increase the cells' replication rate. Those variants which do enhance replication rate will increase in frequency and do so entirely independently of whether the enhanced rate of cell division is of benefit to the individual harboring them. The multicellular individual is an environment potentially populated by both parent and variant daughter cells.

Mammalian cancers are a grim reminder that genetic variants can indeed arise in cell lineages within an organism. The devastating results of the metastasis of cancers are an equally potent reminder that the proliferation of variants can be favored at the level of the cell lineage, quite independent of the consequences to the functioning of the individual harboring the variant.[32] Malignant cells, which may be triggered by as minor a variation as a single base substitution, come to express membrane receptors for the very growth-enhancing factors that they themselves produce. They succeed because mutations giving rise to them ultimately endow them with a greater replication rate than their neighbors.

32. The metastasis of tumors provides a particularly clear instance of selection at the level of the cell lineage. Tumors are well known to become increasingly heterogeneous as they increase in size. Metastatic tumors frequently display considerably less phenotypic diversity, suggesting that the cells of primary tumor have been selected on the basis of their capacity for spread within the somatic environment.

The cells of a metazoan are endowed with the capacity for heritable variation and differential replication—all the components of the evolutionary process. Heritability and variation are not denied to the cell lineages within an organism, nor are cell lineages immune from the rules governing any self-replicating system: variants which enhance replication rate are favored. Heritability, though, has two quite distinct meanings in this context, as August Weismann first made explicit. Variations at the level of the cell lineage may be inherited by lineal descendants within the individual, yet not be passed on to subsequent generations. A cell lineage which becomes terminally committed to some somatic duty is precluded from ever giving rise to gametes—to ever producing a new genetic individual. Heritability *at the level of the individual* is precluded by terminal somatic differentiation. While the cell lineage may not be able to contribute to the production of a new genetic individual, heritability *at the level of the cell lineage* is by no means precluded. The fact that mammalian cancers cannot give rise to gametes in no way limits their replicatory advantage over their neighbors.[33]

The distinction between heritability at the level of the cell lineage and at the level of the individual is a crucial one. Curiously, though, its implications have rarely been explored in depth. Those factors which increase the replication rate of cell lineages within the organism may be deleterious to the individual as a functioning unit, but nevertheless will be inexorably favored at the level of the cell lineage. If a harmoniously functioning unit is to evolve, mechanisms must have evolved whereby variants which enhance their own replication rate by failing to adopt somatic duties are controlled. Selection at the level of the individual must have effectively opposed selection at the level of the cell lineage. The metazoan embryo is hardly immune to this conflict. The ancestral constraint on simultaneous cell division and locomotion demands that the embryo be ciliated. Yet the differentiated cells can no longer divide while in the differentiated state. Clearly the cell lineage producing ciliated products

33. Perhaps the clearest example of the conflict between selection at the level of the cell lineage and that of the individual is that of germ-line cancers. Here enhanced cell division occurs in the very cells which would otherwise be capable of contributing to the next generation. Germ-line cancers represent variants which have abandoned heritability at the level of the individual in return for increased replication of the cell lineage.

will replicate at a lower rate than a comparable cell which does not produce ciliated daughter cells. Why, then, should any cell in a dividing embryo become ciliated, or otherwise differentiated, in a fashion which limits its own capacity for increase? It should not.[34]

The differentiation of some morula cells to become the ciliated cells of a blastula is often the first manifestation of the process. It is, however, only the beginning. The continued development of the embryo involves a cascade of subsequent differentiation events, many of which involve the same conflict. The differentiation of epithelial tissue to presumptive nerve cells signals the entry of these epithelial cells into a pathway in which their own replication rate will ultimately be limited. The neural tissue, upon later contact with epithelial cells, will direct a subset of the epithelial tissue to develop into the lens of the eye, again defining a path of further limitation on cell division. The ontogeny of a metazoan involves a continued sequence in which cell lineages progressively deny their own capacity for increase to the collective interest of the individual. How could an organism evolve such that some cells in that organism abandon their own capacity for replication?

V

Metazoans are manifestly discrete genetic entities, in which cellular differentiation routinely acts to limit the inherent potential of their constituent cells for unbounded growth. The conflict between the potentially opposing processes of somatic differentiation and organismal integrity has been resolved in favor of the individual. The existence of harmoniously functioning multicellular organisms is compelling testimony to this fact. Metazoan evolution is characterized by an increasing sophistication of cells, tissues, and organs which perform somatic duties of value to the individual as a whole, but which require the cells composing them to limit

34. Patterns of cell movement in the early embryo may be profitably viewed in this light. Multipolar introgression, for example, may represent the activity of an ancient variant which favored its own replication by leaving the ciliated blastular wall for the interior, where it might continue to divide and ultimately gain access to the germ line with little or no effect on the survivorship of the primitively unipolar embryo in which it first arose.

ST. CHARLES COUNTY
COMMUNITY COLLEGE LIBRARY
4601 MID RIVERS MALL DRIVE
ST. CHARLES, MO. 63304

their inherent potential for proliferation. The propensity for continued self-replication has been subjugated to the interests of the whole.

The embryos of metazoans have come to control this process in a direct fashion. Their method was demonstrated a century ago. In 1887, Chabry used a crude micromanipulator of his own design and manufacture to destroy cells in the two-cell embryo of the ascidian *Ascidia scabra*.[35] Each altered embryo yielded a half-tadpole. His experiment, closely followed by the divergent results of Roux's similar manipulation of the frog embryo and Driesch's classic work on echinoderms, launched the great experimental reorientation in embryology at the turn of the century. Time and time again, elegant experiments demonstrated that the organization of the unfertilized egg predetermined, to varying extents, the ontogenetic path followed by the embryo.[36] *The developing embryo follows a path of differentiation enforced upon it by its mother.*

Maternally derived determinants, generally in the form of the cytoplasmic elements deposited in the egg, remove control over differentiation from the embryonic cells undergoing that differentiation. The maternal cytoplasm predetermines the pattern of differentiation of early embryonic cell divisions. In its most extreme manifestation, cytoplasmic determinants provided by the mother in the unfertilized egg entirely repress synthesis in the embryo. Alternatively, the

35. Chabry, L. 1887. *J. Anat. Physiol. Paris* 23:167–319.

36. Curiously, it was the very success of these experimental manipulations that sowed the seeds of future misunderstanding between experimental embryologists and transmission geneticists. The chromosome theory of genetics held the tacit assumption that genes were responsible for ontogeny: their faithful replication and equal partitioning were observable in cytological features obvious in any cell independent of its state of differentiation, and mutations occurring in them were expressed in altered developmental pattern. Yet, the experimental manipulation of embryos had repeatedly demonstrated that the nucleus was passive. The nucleus did not direct ontogeny; the temporal and spatial partitioning of cytoplasm did. The chromosome theory seemed at odds with the clear experimental results of embryologists. Both groups had seemingly unambiguous evidence that their own view was correct, and that of their colleagues was, for some obscure reason, misguided. F. R. Lillie (1928. *Science* 66:361–368) states the challenge succinctly, "Those who who desire to make genetics the basis of physiology of development will have to explain how an unchanging complex can direct the course of an ordered developmental system." Lillie's challenge remains today.

embryo may initiate synthesis, but only under the influence of the maternal cytoplasmic agents which activate specific gene sequences in the embryo. Development proceeds, can only proceed, via those instructions left by the mother. When the maternal cytoplasm provides all the mRNA instructions or gene activators necessary to specify those differentiation pathways crucial to the establishment of a functioning organism, the developing embryo can in no way influence its own pattern of differentiation. Selection at the level of the individual has opposed selection at the level of the cell lineage by acting to set the timing of terminal somatic differentiation as far back in ontogeny as possible—wherever possible into the maternal cytoplasm itself.

This simple solution, however, cannot be universally applied. For early development to be directed entirely by maternal instructions, an enormous store of resources will be required for these instructions to be followed. No alternative exists, short of the embryo assuming synthetic tasks itself. Unequal cleavage seems a necessary consequence of the predetermination of embryonic cell fates. Cleavage must result in some differentiated daughter cells and other, as yet undifferentiated, cells. The latter must contain the bulk of the remaining store of maternal instructions and products necessary for their implementation. One subset of cleavage products will necessarily be large, the other small. These considerations frame a dilemma, one shaped on the one hand by a primitive constraint and, on the other, by selective forces acting to determine embryonic cell fates. In forms with external fertilization and free-living young, the blastula must become ciliated. The early cleavage of such an egg must yield cells which serve the same function—they divide to become ciliated. Any gross inequality in the distribution of cytoplasm in such an embryo would result in an embryo which was distinctly unbalanced; the sphere would not be uniformly covered with cilia. Cleavage to produce such a uniformly ciliated surface must necessarily be of cells of roughly similar size, yet control over early embryonic determination demands unequal cleavage.

The simplest of metazoan taxa, the Porifera and the Cnidaria, are characterized by a relatively small number of differentiated cell types and relatively simple construction of cellular and tissue grade respectively. Neither taxa has fully resolved the dilemma: cleavage is equal and cell fate unde-

FIGURE 2.14

Radial cleavage in the
hydroid *Aequorea for-
skalea*. (From Claus,
1883.)

termined until gastrulation occurs in the free-living blastula of the simplest classes of each phylum, the Calcarea and the Hydrozoa (Figure 2.14). The equipotential of products of equal cleavage division is evident in the results of dissection of the 8 (or even 16) cell embryo of the hydroid *Clytia*. Each isolated blastomere forms a fully competent, albeit minute, functional planula larva (Figure. 2.15). Total, equal, and radial cleavage reflects an equivalence of developmental potentials for each of the cleavage products. Even the mature blastula can be halved or quartered to give rise to competent half- or quarter-sized larvae. The hydroid larva is termed regulative; the fate of early cleavage divisions is not terminally established at the onset of cleavage.

In the simplest taxa, extensive maternal direction of somatic cell fate is largely precluded by the need to produce a uniformly ciliated disperal product. Limited unequal cleavage and maternal predestination are not unknown in these phyla, however. They simply do not occur in forms with a free-living morula. Retention of the embryo within maternal tissues to give rise to a free-living blastula momentarily relaxes the constraint on ciliation. Just as maternal brooding up to the point of blastula formation permitted the evolution of different styles of gastrulation, brooding likewise provides an opportunity for some degree of unequal cleavage. The calcareous sponge *Sycon* serves as an example. Initial cleavage divisions are equal such that the embryo, lying within choanocyte chambers of the mother, encircles those maternal cells that guide its development (Figure 2.16). Subsequent divisions are unequal, giving rise to a series of cells which become flagellated. After the repeated iteration of flagellated cells, the embryo abandons its maternal connection, inverts, and swims out into the sea.[37] Unequal cleavage signals pre-

37. An independent phylogenetic experiment in cellular differentiation is found in the colonial flagellate *Volvox*. Much has been made of the phylogenetic significance of the fact that sponge embryos and *Volvox* embryos both display a similar pattern of cleavage and inversion of flagellated cells early in development. The arguments presented here suggest a rather different interpretation. If both taxa are faced with the same constraint on simultaneous cell division and flagellation, then a limited period of brooding offers a temporal respite from the operation of this constraint. This respite provides the potential for unequal cleavage prior to release from maternal tissues. *Volvox* and sponge embryos may display similar patterns in early ontogeny not because they share any common ancestry, but because they face precisely the same functional constraint on their develop-

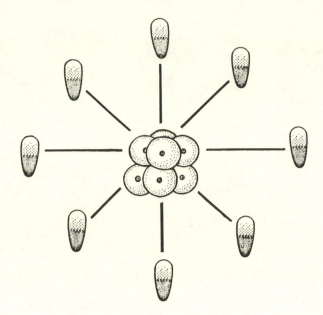

FIGURE 2.15   Developmental fate of isolated blastomeres of the 8-cell embryo of the hydroid *Clytia*.

destination of embryonic fate. Upon transection of the amphiblastula larva into its flagellated and unflagellated halves, the former remains permanently flagellated without further development. These cells have become terminally differentiated. Incapable of locomotion, the larger unflagellated fragments drop to the substratum, but nevertheless continue to develop into quite functional sponges.[38]

ment and have independently evolved the same solution—that of limited brooding permitting unequal early cell divisions.

38. The embryological peculiarities of sponges are one of the central criteria upon which phylogeneticists base their distant separation of sponges from the rest of the metazoa. Aside from the coeloblastula, amphiblastula and parenchymula larva occur. The amphiblastula is typically a hollow sphere constructed of a flagellated and unflagellated pole. The parenchymula larva is typically a solid sphere composed of a ciliated exterior and a diverse, often highly differentiated, interior cell population. The amphiblastula and parenchymula represent increasingly precocious development of adults. Significantly, precocious development is associated with maternal influences on early larval development. In many sponges, including those which display the greatest cellular complexity, the embryo is brooded for relatively long periods. The parenchymula larva of some forms eventually released is composed of most cell types found in the adult.

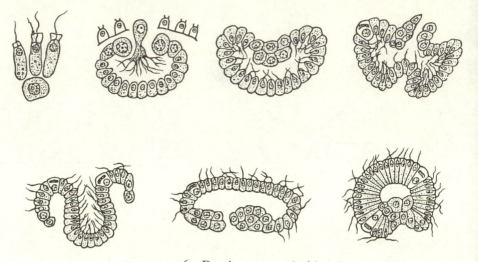

FIGURE 2.16 Development to the blastula stage within maternal tissues of the sponge *Sycon*. The fertilized egg develops under a choanocyte layer of the mother, with maternal tissues involved in the process. After early cleavage divisions, some embryonic cells become flagellated, the embryo inverts, and is released to the sea as an amphiblastula larva. (From Dubosocq and Tuzet, 1942.)

Limited unequal cleavage can occur in organisms with free-living blastula, if the extent of unequal cleavage is not so great as to preclude effective movement. Such is the case in some echinoderms. In echinoids, cleavage is subtly modified from the total radial forms seen in asteroids, ophuroids,

---

Some freshwater forms actually develop miniature flagellated chambers in the larva. Such larvae are little more than miniature adults, in which most of the differentiated cells of the adult have already made their appearance. Viewed from the perspective of brooding as an alternative to a purely cytoplasmic mode of maternal predestination, the embryology of sponges is perhaps not so divergent as is traditionally supposed.

The pattern of increasing maternal control of development via brooding seen in sponges is paralleled in another taxon with otherwise regulative development, the vertebrates. In both groups, the embryo behaves as if it is regulative, but aspects of the normal development of the embryo are clearly under maternal guidance. Although the mechanisms involved in, say, the maternal control of annelid development by cytoplasmic determinants in the egg, and the maternal control of human development, differ in the extreme, the result is the same: embryos in which cell fate departs from the pathway of benefit to the individual may be eliminated in

and holothurians. Cleavage in *Paracentrotus lividus*, for example, is total, equal, and radial through the 8-cell stage. The 16-cell stage, though, sees the cells at the animal pole divided equally, whereas the vegetal pole cells cleave unequally to produce four micromeres and four macromeres. Animal pole blastomeres and vegetal macromeres continue to divide rapidly, come to occupy the bulk of the surface of the blastula, and develop cilia. Vegetal micromeres divide much more slowly, come to occupy only a small surface of the blastula, and eventually enter the blastocoele to form primary mesenchyme. Equal cleavage again signals the production of equipotential cells and unequal cleavage the production of predetermined cells. Isolation of the early equal cleavage divisions give rise to dwarfed but fully competent larvae. Isolation of later unequal stages does not. As in the case of the sponge amphiblastula, isolation of the ciliated animal half of a 32-cell embryo yields a ciliated blastula which remains in that state permanently (Hörstadius' *"dauerblastulae"*), while isolation of some animal pole blastomeres along with micromeres will yield quite normal larvae (Figure 2.17).

Sea urchins have been a favorite object for the study of gene expression in early ontogeny and, in this case, maternal control over this process need not be inferred from the distribution of egg cytoplasm. Here it is known that maternal mRNA directs the ontogeny of echinoids up to the gastrulation stage; virtually no synthetic activity on the part of the embryo itself can be detected prior to this point.[39] The maternal cytoplasm defines which cells will divide equally to become ciliated and which cells will obtain a subset of cytoplasmic determinants allowing them to continue proliferation into the blastocoele. Despite their slightly greater degree of maternal control relative to the free-swimming blastula of sponges and cnidarians, the echinoderm larva is still regulative at early stages; the requirement of a large number of ciliated epithelial cells has limited unequal cleav-

---

both cases. While viviparity is typically interpreted as a reproductive adaptation in the context of selection on individuals for increasing parental investment, the same trait is also favored at the level of the cell lineage in otherwise regulative taxa as a mechanism to control interactions between cell lineages in the developing embryo.

39. Davidson, E. H. *Gene Activity in Early Development*. 2nd ed. New York: Academic Press, 1976.

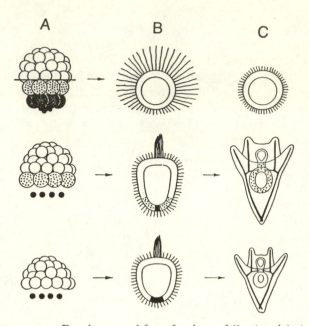

FIGURE 2.17   Developmental fate of embryos following deletion of the entire vegetal pole (A) and of vegetal macromeres (B, C). Note that deletion of vegetal macromeres results in an embryo of near-normal conformation, whereas deletion of vegetal macromeres and micromeres results in a permanently ciliated embryo. (After Hörstadius, 1939.)

age and predetermination to only a small subset of daughter cells. Neither the echinoids, nor the sponges, have fully resolved the conflicting demands of a free-living embryo and simultaneous early terminal differentiation.

Many taxa, though, have resolved the dilemma. They both display a ciliated dispersal stage and are capable of specifying cell fates in the earliest stages of ontogeny. The solutions arrived at in these forms are to be found in the various modifications on the theme of gastrulation outlined earlier. In forms in which development proceeds through gastrulation within maternal tissues or within egg capsules, the free-swimming stage, and hence the stage requiring ciliation, has been shifted in ontogeny from the blastula to some period post-gastrulation. Here the original constraint is relaxed and selection at the level of the individual would be expected to yield unequal cleavage and highly determined cell fates.

Consider the turbellarian flatworms as an example. Turbellarians occupy a central phyletic position. As the simplest bilateral organisms, they display the first appearance of several cell types and the first organization of some differentiated cells into discrete organ systems. Associated with this great cellular diversification is the first appearance of a mode of cleavage shared by many higher coelomate groups. The typical spiralian cleavage appears (Figure 2.18). Each quadrant of the fertilized egg initially cleaves to produce two cells of unequal size. The smaller cell then undergoes a second cleavage to produce yet smaller cells which rest within the cleavage furrows of the first. The third cleavage again results in a reduction of cell size, with each resulting micromere again lying within the cleavage furrow of the last. A typical spiral pattern is formed (Figure 2.19). The blastula is composed of cells of unequal size, and gastrulation occurs by the proliferation of micromeres to enclose the far larger macromeres. This is epiboly, the alternative mode of gastrulation found in those taxa which retain or encapsulate their embryos until hatching late in development.[40]

Each early cell division in the typical spiral cleavage involves the somatic determination of that cell and its daughters. For example, in the annelid, *Sabellaria vulgaris*, separation of the blastomeres at the 2-cell stage results in only the larger cell developing into a normal trochophore larva. The smaller blastomere will continue to develop but will lack essential organs. No less striking results are obtained with molluscs, crustaceans, and ctenophores: isolation of the unequal first cleavage typically results in one functional, albeit misproportioned, larva and one dysfunctional fragment. Isolation at later stages produces even more extreme results. Witness the fate of isolated blastomeres of the 16-cell *Nereis* embryo (Figure 2.20), in which the isolates yield only those portions of the larva that they would have had if they had not been isolated. In contrast to regulative embryos, such as the 16-cell hydroid embryo which can be manipulated to produce sixteen functional larvae, these embryos are mosaic: the fate of cells in the early embryo is established at the onset of ontogeny.

40. Uncontrolled malignant growth may be triggered by the expression of receptors for specific growth factors. But just as critical may be the removal of inhibitors which under normal conditions act to restrain cell growth.

FIGURE 2.18

Isolation of one quadrant of the spirally cleaving egg of the mollusc *Trochus*, illustrating the progressive decrease in cell size. (From Kume and Dan, 1957.)

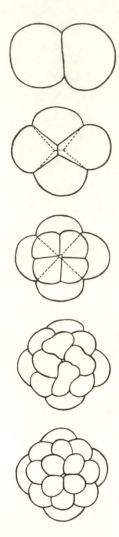

FIGURE 2.19

Spiral cleavage in poly-
clad flatworm *Noto-
plana humilis*. (From
Kato, 1940.)

The influence of maternal cytoplasmic control of early ontogeny in these forms is legion. Late-nineteenth-century experimental embryologists charted the path of pigmented cytoplasm in unfertilized eggs to the respective regions in the
embryo, and performed manipulations to demonstrate the
role of maternal cytoplasm as morphogenetic determinants.
Of the many demonstrations of the influence of unequal
cleavage enforcing maternal cytoplasmic control, none is
more striking than that found in many molluscs and certain
annelid embryos. Here the doling out of cytoplasmic determinants during early development is observed in a unique
morphological feature. At the time of the first cleavage division a pronounced cytoplasmic protuberance is formed,
called the polar lobe. The lobe, which can be as large or
larger than the blastomeres themselves, is not a cell, but
merely a sac harboring maternal cytoplasmic contributions.
The first cleavage is roughly equal, but cleavage is immediately followed by the resorption of the polar lobe by one cell,
yielding two cells of highly unequal size. The lobe reappears
at the second equal cleavage, only to be absorbed again to
produce dissimilar cell sizes. This process of equal cleavage
followed by absorption of the polar lobe to generate unequal
cell sizes and distributions of maternal determinants may
continue for up to eight division cycles. The polar lobe invites manipulation to restore equal cell size (Figure 2.21).
Deletion of the lobe, predictably, produces highly aberrant,
dysfunctional larvae. Compression of the embryo at the first
cleavage division has equally striking effects. Compression
forces the contents of the polar lobe into both cleavage products, mimicking the effects of an equal cell division (Figure
2.22). Compression yields a doubled larva, illustrating in
dramatic fashion that unequal cleavage and segregation of
cytoplasmic determinants produce the very patterns in determination at issue.

The pattern of abandoning the free-living blastula to produce a free-living dispersal stage after gastrulation is found
in many groups that are distantly, if at all, related. Early ontogenies in many higher protostome phylum (the annelids,
molluscs, and arthropods), and in the radiate phyla Ctenophora, have this feature in common. The embryo becomes
ciliated, but not until after gastrulation is complete. The
timing of the ciliated dispersal stage has undergone a heterochronic shift from the blastula stage to post-gastrulation,

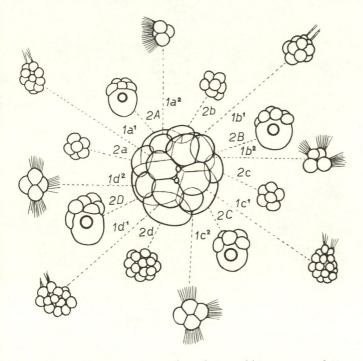

FIGURE 2.20 Developmental fate of isolated blastomeres of the 16-cell embryo of the annelid *Nereis*. (From Costello, 1945.)

releasing such taxa from the ciliation constraint. The demand for equal cell divisions is relaxed. In each case, either the spiralian pattern of cleavage has arisen or another idiosyncratic mode of cleavage is evident, and mosaicism has arisen in all. Selection at the level of the individual has enforced early somatic determination convergently in all these groups.

Even more extreme modifications are found in those forms which have abandoned the free-swimming dispersal stage entirely. The demands of somatic determination and free-swimming dispersal are no longer in conflict. Here cleavage need not be equal at early stages to produce a ciliated blastula, nor must the benthic adult pass through a specialized dispersal stage. The zygote can give rise directly to an adult. Trematodes, nematodes, rotifers, and tardigrades are all sophisticated organisms displaying considerable cellular complexity, and all develop in a peculiar fashion. All of them de-

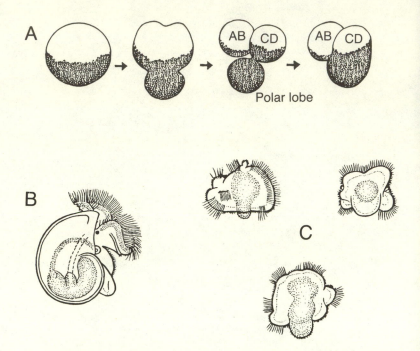

FIGURE 2.21   The first cleavage division in the snail *Ilyanassa*, showing the formation and resorption of the polar lobe (A). In contrast to normal development (B), deletion of the lobe at this stage results in highly aberrant, dysfunctional larvae (C). (Part (A) from Raff and Kaufman, 1983; (B–C) from Clement, 1952.)

velop within egg capsules, cleave in unusual, taxon-specific ways, and hatch as miniature adults. The development of nematode *Ascaris* is illustrative (Figure 2.23). Here the embryonic fate of all subsequent cells is determined within the first four cleavage divisions. The first cleavage gives rise to an uncommitted cell and the first somablast. The latter is already terminally committed to eventually form the primary ectoderm of the animal; the former divides to produce another undifferentiated daughter and the secondary somablast, which is, in turn, terminally determined to form the endoderm and primary mesoderm. Within two subsequent divisions of the undifferentiated lineage, primordia of all somatic lineages of the adult have become established, leaving only primordial germ cells as undifferentiated cells which give rise to gametes. This entire process is under strict maternal control; synthetic activity on the part of the embryo occurs only after somatic determination has become established. Maternal mRNA instructs embryonic development until the final determinative cleavage giving rise to germ cells. Total maternal determination of the somatic cell fate of the subsequent generation is realized in these forms. An individual cannot influence the state of differentiation of its own cells by products of its own making. The predicted pattern has been realized—the timing of terminal determination of embryonic cell fates has been pushed back as far into early ontogeny as possible.

Just as the demand for simultaneous ciliation and continued development gave rise to gastrulation, the demand for maternal control of differentiation gave rise to patterns in cleavage and regulation. The study of cleavage and regulation is the study of the timing of terminal somatic differentiation. Control over this process—that is, control of selection at the level of the individual over selection at the level of the cell lineage—could proceed only as far as is allowed by the sequential improvements made in response to the ancestral constraint. The primitive ciliated blastula demanded gastrulation, but limited early determination. Relaxation of this requirement led to alternative modes of gastrulation and yet earlier determination. Finally, removal of the requirement allowed total control of cell fate and, in extreme cases, the abandonment of gastrulation itself. Early embryonic development, the complex interplay of cell division, cell differentiation, and morphogenetic movement, is the joint consequence of ancestral constraints mediating the interaction between selection at the level of the individual and selection at the level of the cell lineage.

VI

The synthetic theory of evolution, with its emphasis on the individual as the unit of evolutionary modification, is frequently, and justly, criticized as a "theory of adults"—one which has failed to address the diversity of ontogeny. Evolutionists, even today, seek to understand how development will illuminate patterns in evolution, not how evolution will illuminate the details of the developmental process. To some considerable degree, the lack of ontogenetic theory is more a matter of benign neglect of development by evolutionary theorists, than one of inherent limitation in the current lan-

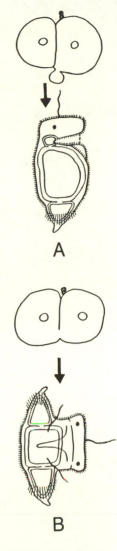

FIGURE 2.22

Normal development of the polychaete *Chaetopterus* (A) and the production of a twinned larva (B) by compression of the polar lobe at first cleavage to distribute polar lobe material equally between the two blastomeres. (After Titlebaum, 1928.)

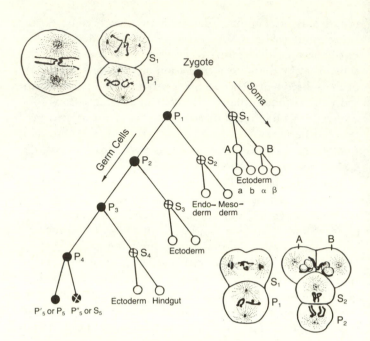

FIGURE 2.23   Schematic diagram of the early cleavage divisions and resulting determination of cell fate in the nematode *Ascaris*. (After Nieuwkoop and Sutasurya, 1981.)

guage of evolutionary theory. Some patterns in metazoan development may be approached by quite standard neo-Darwinian cognitive styles. The hypothesized relationship between gastrulation and the interplay of an ancient constraint with the need for a free-swimming larval stage is little more than a suggestion that a particular adaptation, gastrulation, was shaped by selection on the embryo for a given function. The pattern of gastrulation, and its modification and abandonment when selection is removed, is quite typical adaptationist thinking.

Selection at the level of the individual, however successful in defining why gastrulation evolved and why it displays the phyletic patterns that it does, offers little insight into many parallel problems. The diverse routes of cleavage and regulation by which embryos arrive at gastrulation are in no fundamental way illuminated by arguments focused on the level of the individual. Individual selection, *a priori*, is indifferent to whether a somatic cell is determined in the maternal cytoplasm or in later stages in development, just as long as it arises in time to perform its particular somatic duty.[41] Selection at the level of the individual would certainly demand that somatic duties be performed, but why should the determination of these patterns so uniformly occur in earliest ontogeny, even before the tissues in which they are ultimately expressed face any interaction with environmental challenges? Even more curiously, why should the extent of predetermination reflect the ultimate cellular complexity of the taxa in which it occurs, and why should these very patterns in cleavage and regulation display a clear and compelling phyletic pattern that mirrors the alternative modes of gastrulation permitted by the original constraint?

Patterns in cleavage and regulation become intelligible only upon consideration of selection at the level of the cell lineage. A failure to predetermine differentiation pathways produces an individual potentially susceptible to behavior on the part of its constituent cell lineages which favor their own replication at the expense of the viability of the individual. With the recognition that patterns of differentiation represent selection against the unbounded replication of cell lineages within the organism, the matter is considerably clarified. Now the issue of the timing of maternal determination has a clear basis in the interplay between selection at the level of the cell lineage and the level of the individual. Differentiation limits clonal iteration of cells within the organism and its occurrence must be enforced. Neo-Darwinian cognitive style can now be brought into play; patterns in cleavage and regulation are adaptations that serve the function of imposing selection at the level of the cell lineage.

A fundamental thesis emerges: the evolution of individuality becomes comprehensible not by the study of ancient constraint, nor by the study of selection upon individuals, nor even by the study of selection on cell lineages, but only through the study of their interaction. If an original con-

41. One might argue that early determination reflects selection at the level of the individual for rapid development. While this is likely a contributing selective force operating at the level of the individual, this explanation alone is insufficient. Plants, for example, are not challenged by the ciliation constraint discussed here, but are presumably under similar selection for rapid development. Yet plants do not display the elaborate mechanisms of maternal predestination of cell fate seen in animal embryos.

straint could not be specified, there would be no point of departure—any ontogenetic pattern would be conceivable. If a selective force acting on entire individuals were not known, there would be no basis for assigning a function to gastrulation, nor hypotheses for why some organisms gastrulate and others do not. If the role of differentiation as a selective force acting on cell lineages were not articulated, there would be no grounds for expecting maternal predetermination or basis for predicting its differential extent in different taxa. The conservatism of early ontogeny, expressed as the various modes of cell division, cell differentiation, and morphogenetic movement, become penetrable to standard neo-Darwinian thinking with one simple modification: evolutionary pattern has arisen not by selection on individuals alone, but by the interactive effects of selection operating at differing levels of biological organization.

THE DIVERSITY OF LATE ONTOGENY

I

After gastrulation, the initially stereotyped development of metazoan embryos disappears into complex, taxon-specific modes of elaboration from which one of several discrete body plans emerges. These fundamental "bauplans" evolved rapidly and, inexplicably, became fixed. The first unicellular eukaryotes are known from the Bitter Springs Formation (0.9 b.y.b.p.); yet metazoans clearly recognizable as members of extant phyla are found only 200 million years later in the Ediacara Formation (0.7 b.y.b.p.). The fundamental body plans of extant taxa swam, floated, and crawled in the early Paleozoic oceans. Modifications upon existing plans certainly arose: land was conquered, flight evolved, and myriad species-specific complexities ramified. Yet, between the Cambrian (0.6 b.y.b.p.) and the Recent, no new body plan has arisen. In over half a billion years, metazoan development has followed the same basic formulae. The major innovations of metazoan development were experiments of the Precambrian.

A paleontological preamble casts the issue in a dual light: we must not only seek to discover the processes shaping the evolution of development, we must also discover the forces which have so effectively restrained their further modification. Did the forces shaping evolution change between Bitter Springs and Ediacaran times? In an important sense, they did. The eukaryotes of Bitter Springs were single cells or simple colonies of monomorphic cells. Each cell had the capacity to form a new lineage; selection on the individual and selection on the cell lineage were equivalent. By the Ediacaran period, metazoans had evolved, and some cells had already lost their capacity to form a new individual. Selection acted not only upon cells, but also upon multicellular individuals. During this crucial period, the cells of metazoans diverged from unicellular eukaryotes in the all-important sense that heritability, the capacity to yield a new multicellular individual, became restricted to a subset of cell lineages. Evolutionary history literally entered a new phase. No longer did self-replication alone ensure heritability.

II

The path from a unicellular condition to a multicellular one has been well-traveled. Of the some 23 monophyletic protist groups, fully 17 have multicellular representatives. The path from multicellularity to cellular differentiation, however, proved a far less porous filter. Of the 17 multicellular taxa, only 3 groups—the plants, the fungi, and the animals—have developed cellular differentiation in more than a handful of species. With the evolution of cellular differentiation, kingdoms were made of some protist groups; yet we know virtually nothing as to why this transition was closed to all but a few taxa.

Minor taxa, which possess only a small number of species with cellular differentiation, provide valuable clues to the roadblocks in the path to the differentiated state; these clues are easily obscured in more complex taxa. The slime molds are one such group, fascinating for their curious mode of development and well characterized as a laboratory model of a simple cellular differentiating system. These free-living amoeboid or flagellate cells aggregate in the face of adverse environmental conditions to form a multicellular fruiting body. Cellular differentiation has arisen in the two quite distinct slime mold clades, the acrasids and the dictyostelids. For example, in the acrasid *Guttulinopsis*, the fruiting stage of the life cycle amounts to little more than a mound of cells, the majority of which are totipotent to yield new clones, whereas in the related genus, *Acrasis*, a complex branching structure is formed in which some cells are terminally committed to the purely somatic duty of producing a stalk (Figure 3.1). The transition from multicellularity to cellular differentiation is mirrored in the dictyostelid slime molds. The fruiting body of the dictyostelid *Acytostelium* is acellular and all the cells of an aggregation are totipotent, but in the related genus *Dictyostelium*, the cells of the stalk are differentiated like those of *Acrasis* (Figure 3.1).

The differentiated state, in which an aggregate gives rise to somatic stalk cells and totipotent spore cells, is not an entirely stable condition; rather it routinely reverts to a multicellular state. In 1962, M. Filosa, working with a standard

A

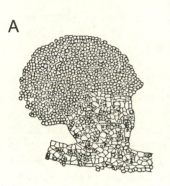

B

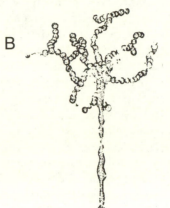

FIGURE 3.1

Structure of the acrasids *Guttulinopsis vulgaris* (A) and *Acrasis rosea* (B) and the dictyostelids *Acytostelium leptosomum* (C) and *Dictyostelium discoideum* (D). (Parts (A) and (D) from Olive, 1975; (B–C) from Bonner, 1967.)

C

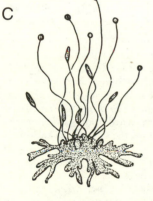

D

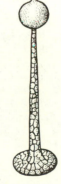

laboratory strain of *Dictyostelium mucoroides*, found that spores isolated from a single sorus produced slime molds of quite different morphologies.[1] The fruiting body was chimeric, yielding a variant line which produced no stalk cells whatsoever. The variant line divided nicely, aggregated nicely, and produced perfectly viable spores. It had simply reverted to the multicellular condition (Figure 3.2). Subsequent studies have shown that phenotypically similar variants arise in natural populations and are capable of persisting in dynamic equilibrium with normal strains.[2]

An independent phylogenetic experiment in cellular differentiation is found in the Volvacales, a favorite model for the evolution of cellular differentiation since Weismann's time.[3] In the simpler species of these colonial flagellates (e.g., *Gonium, Pandorina*), a multicellular colony occurs in which all cells are totipotent to produce new colonies; there is no hint of cellular differentiation. True cellular differentiation, however, does occur in the genus *Volvox* (Figure 3.3). Of the several thousand cells in these spherical colonies, only a few are capable of generating daughter colonies. The remainder become flagellated and are incapable of further division. Just as the slime mold *Dictyostelium* displays spontaneous reversion to the multicellular state from one of cellular differentiation, so does *Volvox*. Mutants of *Volvox* are

1. Filosa, M. F. 1962. *Am. Nat.* 96:79–91.
2. Buss, L. W. 1982. *Proc. Nat. Acad. Sci. USA* 79:5337–5341.
3. The appearance of early embryonic differentiation of the germ line in *Volvox* was a powerful element in Weismann's arsenal, as he saw this organism as the ancestral stock of metazoans. He points out that it is important ". . . to ascertain whether here, at the very origin of heteroplastid organisms, the germ-cells, sexually differentiated or not, arise from the somatic cells *at the end of ontogeny*, or whether the substance of the parent germ-cell, during embryonic development, is *from the first* separated into somatic and germ-cells. . . . The germ-cells of *Volvox* are differentiated during embryonic development, that is, before the escape of the young heteroplastid organism from the egg capsule. We cannot therefore imagine that the phyletic development of the first heteroplastid organism took place in a manner different from that which I have previously advocated on theoretical grounds, before this striking instance occurred to me." (Italics in the original. Weismann, A. *Essays on Heredity and Kindred Biological Problems*. E. B. Poulton, S. Schonland, and A. E. Shipley, eds. and trans. Oxford: Clarendon Press, 1889:204.) Weismann's interpretation, of course, proved erroneous. Germ cells are differentiated at the end of ontogeny in the most primitive metazoan taxa (see Table 1.1).

FIGURE 3.2   The development of (A) wild-type *Dictyostelium mu-*
coroides and of (B) a variant that produces little or
no somatic tissues. (After Bonner, 1967.)

known in which cellular differentiation is lost; all cells are
capable of generating a new colony.[4] In a single step, cellular
differentiation has been lost and *Volvox* has returned to its initially multicellular state.

The slime mold and Volvacales are illustrative of a general
phenomenon. Attendant with the evolution of cellular differentiation came a previously unknown selective scenario.
In a multicellular organism lacking cellular differentiation,
all cells are potentially capable of generating a new colony,
whereas in a similar form with cellular differentiation, some
cells are necessarily relegated to somatic duties. When cellular differentiation first arose, heritability was denied to
some lineages. The new vista is straightforward: a mutant
cell line within an organism may arise which does not relin-
quish heritability.

4. Kochert, G. In C. C. Markert and J. Papaconstantinou, eds. *Develop-*
mental Biology of Reproduction. New York: Academic Press, 1975:55–90.

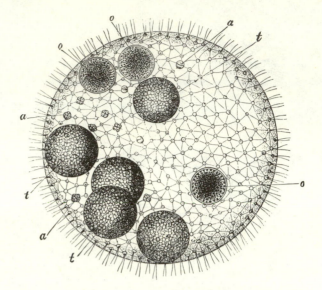

FIGURE 3.3   *Volvox aureus*, showing flagellated somatic cells coating the exterior of the sphere, male germ cells (a), fertilized eggs (o), and several large parthenogenetic egg-cells (t). (From Weismann, 1904.)

The heritability of variants arising in the course of ontogeny is not a process lost forever in the evolutionary past. Transformations of stem or somatic cells to germinative cells still occur in some metazoans. Fully differentiated choanocytes of sponges, which perform the somatic functions of water transport and feeding, are capable of differentiating into sperm or eggs.[5] In fact, entire choanocyte chambers in *Hippospongia communis* are known to be transformed into pockets of spermatogonia (Figure 3.4). Somatic transformations are not limited exclusively to sponges.[6] Experimental

5. Tuzet, O., R. Garrone, and M. Pavans de Ceccatty, 1970a. *C. R. Acad. Sci. Paris* 270:955–957; *Ibid.* 1970b. *Ann. Sci. Nat. Zool. Biol. Anim.* 12:27–50; Diaz, J. P., R. Connes, and J. Paris. 1973. *C. R. Acad. Sci. Paris* 277:661–664; *Ibid.* 1975. *J. Microsc.* 24:105–116.

6. For most somatic cell types, it is not known whether differentiation is reversible. Although some specialized cell lineages of metazoans are certainly terminally differentiated (due, for example, to the distribution of cytoskeletal features which preclude further cell division, or to a loss of nuclear material), the state of differentiation of many cell lineages may be actively maintained by surrounding cells. Cases of annelid and flatworm germ cells appearing *de novo* subsequent to experimental transection, as well as several classical embryological models such as lens regeneration,

removal of gonadal segments in the annelids *Lumbricillus* and *Perionyx*, a process which may be mimicked in nature by predators, leads to *de novo* formation of primordial germ cells from ordinary, otherwise somatic, parietopleural cells.[7] These cases are but the remnants of a process once rampant.

Variants arising in the course of ontogeny are also heritable if they occur in the stem cell lineages giving rise to the germ line. One or more cell lineages remain competent to form the gametes throughout ontogeny in the Porifera, the Cnidaria, the acoel Platyhelminthes, and the Bryozoa. In

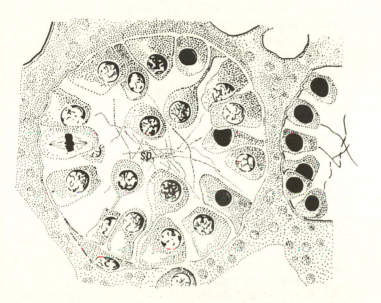

FIGURE 3.4   Cross-section through a choanocyte chamber of the sponge *Hippospongia communis* displaying developing spermatogonia (sp). (From Tuzet, 1964.)

testify to the occurrence of active maintenance of differentiation by surrounding cells. A cell lineage may *become* differentiated as a consequence of its own prior history of gene expression, but it may *remain* differentiated as a consequence of the activities of surrounding cells. Both the literature on regeneration and the literature on metamorphosis might profitably be mined for potential models of differentiation actively maintained by external controls.

7. Herlant-Meewis, H. 1946. *Arch. Biol.* 57:197–306; Gates, G. E. 1943. *Proc. Nat. Acad. Sci. India* B 13:168–179; *Ibid.* 1951. *Proc. Nat. Acad. Sci. India* 34:115–147.

many other metazoan taxa, including some annelids, molluscs, ascidians, echinoderms, arthropods, and even some vertebrates, the germ cells only become recognizable after the major organ systems of the adult have become determined, with reports of their origin being attributed to several distinct cell populations (Table 1.1).[8] Variants arising *in the course of ontogeny* may prove heritable.

The fate of any variant arising in the course of ontogeny will be determined not solely by its interaction with the external environmental milieu, but also by interactions *within* the organism. The organism is an environment potentially populated by normal and variant cells. The variant must compete with the lineage from which it arose for growth-enhancing substances and, ultimately, for access to the germ line.[9] Competition of variants arising within the ancestral metazoan must have held both opportunity and threat—opportunity in the sense that variants which favor both their own replication and that of the organism harboring them may arise and prove heritable.[10] This new scenario provides a potent mechanism for the origin of novel interactions between cell lineages in the course of ontogeny—a source of novel epigenetic interactions. More frequently, however, the heritability of variants must prove a threat. Any variant which reduces its contribution to somatic cell functions, as in the case of *Volvox* and slime molds, would increase dispro-

8. Nieuwkoop, P. D. and L. A. Sutasurya. *Primordial Germ Cells in the Invertebrates*. Cambridge University Press, 1981; Buss, L. W. 1983. *Proc. Nat. Acad. Sci. USA* 80:1387–1391.

9. Competition of this sort is demonstrably common (Buss, L. W., in J.B.C. Jackson, L. W. Buss, and R. E. Cook, eds. *Population Biology and Evolution of Clonal Organisms*. New Haven, Conn.: Yale University Press, 1985:467–505).

10. Although incompletely documented, the first example of the inheritance of a beneficial variant may be found in Weismann's reactions to criticisms of his doctrine of the continuity of the germ plasm. Detmer, a botanist critical of the doctrine, cites as evidence the events following exportation of edible cherries from England to Ceylon. Cherries in temperate climates are deciduous and edible cherries were then, as now, propagated by grafting. Shortly after their importation to Ceylon, some grafts became evergreen, as they have remained since. The parsimonious explanation is an evergreen variant arose in the totipotent meristematic lineage of the graft and, by virtue of a clear competitive superiority, became the dominant meristematic cell line, passing on the trait to their descendants. Unfortunately, Detmer interpreted this phenomenon as a demonstration of the inheritance of acquired modifications and, quite correctly, Weismann refused to defer.

portionately in frequency compared with cell lineages which contribute to somatic duties. Selection at the level of the cell lineage favors such variants and sets this type of selection firmly in opposition to selection at the level of the individual. The heritability of variants arising within the course of ontogeny holds both promise and the potential for devastating conflict.

The perspective of the Modern Synthesis invites the view that selection acts upon individuals, not within them. This view is largely vindicated for many organisms that surround us today (Table 1.1). Yet, nearly a billion years ago, when cellular differentiation first arose, this was assuredly not the case. In those simple cellular differentiating systems, variants arising during the course of ontogeny—within a single organism—must have been first selected within the somatic environment. Conflicts and synergisms between selection at the level of the individual and selection at the level of the cell lineage could not have failed to challenge the ancestral metazoan, nor have failed to imprint themselves upon the modes of development the metazoans ultimately adopted.[11]

III

The growth and interaction of the clonal lineages arising from a metazoan zygote are a necessary consequence of their genetic heritage. Metazoans inherited from their protistan ancestors a complete machinery for cell division. The genetic code, the systems of transcription and translation, and the cytoplasmic architecture to deploy myriad gene products were all actors in the evolutionary play long before metazoans arose. The cells of a developing metazoan embryo follow the same rules governing any self-replicating system:

11. The examples of the simple cellular differentiating systems of slime molds and *Volvox*, in which detrimental variants reversing the state of differentiation are known, should not be taken to imply that either group is without devices to limit the effects of such variants. The slime molds, for example, are largely immune to variants arising in the course of ontogeny by virtue of their habit of living as independent, free-living cells for all but a brief interval of the life cycle. *Volvox* has also evolved traits which limit the heritability of variants. *Volvox* has arrived, presumably via convergence, at the same solution as have most higher protostomes and deuterostomes. Their primordial germ cells are determined in very early ontogeny, limiting heritability to the products of only a few cell divisions per generation.

they divide until some external force limits their further replication. The metazoan innovation is the evolution of epigenetic controls on the growth of developing cell lines, which provide restraints on their inherent propensity for self-replication in a precise cascade such that cell lineages organize themselves into a functional bauplan.

The fact that the diversity of late metazoan development proceeds as a complex of interactions between cell lineages, each individually endowed with a genetic architecture competent to divide without restraint, invites the interpretation that *metazoan development reflects a legacy of past interactions between variant cell lineages arising within the ontogeny of ancestral forms*. Variants which further their own replication rate by restraining or directing the activities of neighboring cells may conceivably simultaneously enhance their own replication and the survivorship of the individual harboring them. Such variants, if they prove heritable, would produce an individual whose offspring would replay, as an epigenetic phenomena, the same sequence of interactions between developing cell lineages.[12] Repetition of this process for a sequence of different variants and corresponding tissue interactions would yield an ontogeny which unfolded as a complex pattern of epigenesis (Figure 3.5). Metazoan development today is manifestly a process of the sequential origin of, and interaction between, cell lineages arising in the clonal progression from the zygote. The fact that metazoans develop via a complex of epigenetic interactions between cell lineages is *prima facie* evidence that the principal modes of metazoan development arose as variants in the course of ontogeny.

While the occurrence of variant cell lineages which, in furthering their own replication, produced innovations of advantage to the individual harboring them must have been infrequent events, such variants are hardly inconceivable. The ancestral metazoan, frequently equated with a spherical colony of choanoflagellates, was a pelagic organism. The origin of a variant cell line which entered the center of such a sphere to continue cell division, similar to that seen in the

12. The interpretation of any particular development phenomenon as an ancient interaction between variant cell lineages is subject to the same restrictions as any other argument involving recapitulation. For a thorough discussion of the limits of recapitulatory reasoning, see Gould, S. J. *Ontogeny and Phylogeny*. Cambridge, Mass.: Harvard University Press, 1977.

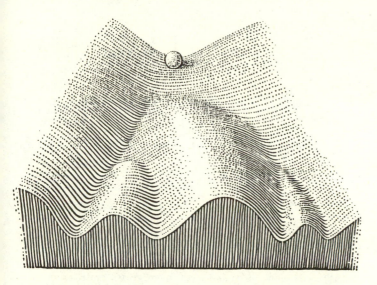

FIGURE 3.5   An epigenetic landscape, Waddington's representation of epigenesis as valleys in a state space. In the view advocated here, each valley represents the origin of a variant cell lineage in the course of ontogeny which effectively competes with existing lineages to establish a new epigenetic progression. (From Bonner, 1958.)

choanoflagellate *Proterospongia*, may have produced a structure which was sufficiently negatively buoyant to fall to the sea floor. Many modern sponges, the taxa most reminiscent of choanoflagellates, do just this. A flagellated sphere populated by amoeboid cells simply drops to the ocean bottom. Indeed, some zoologists regard the enigmatic, crawling placozoan *Trichoplax* as a "living fossil" of this grade of organization, imtermediate between colonial flagellates and sponges (Figure 3.6).[13] The pelago-benthic life cycle of sponges may have arisen as a consequence of variants which, in pursuing their own replication, fortuitously presented the individual with a benthic existence and all the attendant opportunities inherent in the invasion of a new adaptive zone.

Similar scenarios are easily imagined for the origin of other developmental processes among metazoans. The first bilateral animal, presumed by most zoologists to be a planuloid organism, is most closely approximated in the acoel

13. Salvini-Plawen, L. 1978. *Z. zool. Syst. Evolut.-forsch.* 16:40–88.

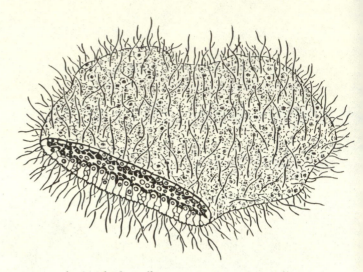

FIGURE 3.6 *Trichoplax adhaerens*, an enigmatic crawling meta-zoan of simple organization. (From Margulis, 1982.)

flatworms. Some acoels lack definitive muscle cells. Rather these worms display a composite epithelio-muscular cell, a cell type characteristic of the Cnidaria. Muscle fibers occur at the base of otherwise typical epithelial cells. In higher tur-bellarian flatworms, however, definitive muscle cells and ep-ithelial cells occur. In these same acoels, epithelial cells may also contribute to the cellular content of the mesenchymal matrix. From such mesenchymal cells the gametes are de-rived. Variants arising in the epithelium, gaining access to the mesenchyme, escape from a purely somatic fate and re-main in the lottery of potential germ cells. In so doing, these cells may carry with them any variants which arose in the on-togeny of the epithelium. An epithelio-muscular cell which abandoned the somatic duty of producing muscle fibers in its base, might not only continue to divide to produce an amus-cular epithelial cell; this same variant might gain access to the germ line to produce descendants with a similar habit.[14]

14. In addition to the primitive character of their musculature, many acoels lack definite organs of reproduction, excretion, sensation, or diges-tion. Rather, in various species the nervous system is a diffuse nerve net, the gut a simple sac, and genital pores entirely lacking. In the higher flat-worms, fully developed organs of reproduction, excretion, sensation, musculature, and digestion appear, as may be expected to be fueled by the

Excercises in theoretical zoology, such as the cartoons proffered above for sponges and flatworms, no doubt lie, in Libbie Hyman's characteristically bald terms, "in the realm of fantasy." They are introduced here not as literal historical reconstructions, but rather as exercises demonstrating the inherent plausibility of a particular perspective on metazoan development. Variants may arise in the course of ontogeny which simultaneously favor their own replication and that of the individual harboring them. If these variants find their way to the germ line, they may effectively establish an epigenetic program. In the myriad details of metazoan development is written a partial record of the interactions between variant cell lineages, particular sequences of which gained access to the germ line to be revealed today as the epigenetic programs by which metazoan embryos developed.

*

Gould and Lewontin recount an anecdote arising from a visit by Herbert Spencer to a laboratory studying fingerprints:[15]

> He asked me to show him my Laboratory and to take his prints, which I did. Then I spoke of the failure to discover the [evolutionary] origin of these patterns, and how the fingers of unborn children had been dissected to ascertain their earliest stages, and so forth. Spencer remarked that this was beginning in the wrong way; that I ought to consider the purpose the ridges had to fulfill, and to work backwards. Here, he said, it was obvious that the delicate mouths of the sudorific glands required the protection given to them by the ridges on either side of them, and therefrom he elaborated a consistent and ingenious hypothesis at great length. I replied that his arguments were beautiful and deserved to be true, but it happened that the mouths of the ducts did not run in the valleys between the crests, but along the crests of the ridges themselves.[16]

---

adoption of a bilateral habit. It is striking, and likely not coincidental, that the fundamental innovations of metazoan organ systems of both ectodermal and endodermal origin first appear in the very taxon in which both ectodermal and endodermal cell lines have access to the germ line.

15. Gould, S. J. and R. C. Lewontin. 1978. *Proc. Roy. Soc. Lond.* 205:581–598.
16. Galton, F. *Memories of My Life.* London: Methuen, 1909:257.

The story is telling. Few evolutionists have not played the role of Spencer in modern reenactments of similar dialogue, as assured as was Spencer in the wisdom of their reasoning. And few modern reductionists have not felt frustration over their colleagues' confusion of that assurance with established empirical fact.

The perspective on the evolution of development as the legacy of past competition between variant tissues to further their own replication, with the fortuitous effect of establishing and subsequently stabilizing an epigenetic program, is an internally self-consistent extension of first principles. Nature, however, is inherently devious and any such theory lacks authority without a demonstration that the genome actually does direct ontogeny in the fashion suggested. The metazoan genome need not, indeed cannot, specify each and every detail of development; metazoan cells divide and interact as a necessary consequence of traits already developed by their ancestors. For ontogeny to proceed, the genome need only encode the relative competitive mechanisms of cell lineages. Epigenetic programs, when unraveled at the molecular level, should reveal that the genome produces ontogenetic pattern solely by directing the relative competitive relationships of developing cell lineages.

Normal development, if this view is to be vindicated, should reflect a complex of interactions in which a cell lineage participates in a function of benefit to the organism as a whole only if participation in that function increases its own rate of replication or if participation in that function is actively enforced by a competing cell lineage. Just as normal somatic function should occur when participation in somatic function favors the replication of the cell lineage participating in that function, mutants which disrupt normal development may be expected to involve lineages which increase their own rate of replication at the expense of the individual as a unit. Consider, for example, the development of neuronal tissue from the embryonic ectoderm in *Drosophila*. In normal development, neuroblasts arise from some cells of the presumptive neuroepithelium and epithelial tissue from others. In the *Notch* mutation, neuroblasts are overproduced at the expense of epithelial cells. Sequencing of an entire 10.5 kb gene of *Notch* has recently revealed that the gene includes no fewer than 36 tandem repeats of a sequence homologous to the epidermal growth factor (EGF).[17] The suggestion inherent in this sequence is unmistakable: the *Notch* gene, by virtue of its EGF sequences, encodes for a greater competitive ability of neuroblasts relative to that of presumptive epithelial cells. The variant, in favoring its own replication at the expense of the individual, is lethal.[18]

In contrast to the case of the *Notch* mutation, normal function is expected to arise from a nexus of interactions between cell lineages in which each cell lineage enhances its own replication by participation in that function. No clearer example of organismal function arising through epigenetic interactions can be found than that of the immune system of mammals. The immune system displays a phenomenal capacity to recognize foreign substances, to free the bloodstream of them, and to accomplish its diversity of function while simultaneously avoiding healthy cells of the same body. The specificity of recognition is of legendary proportions; humans are said to be capable of recognizing up to one billion different foreign substances. No clearer challenge can be made to the suggestion that the genome orchestrates each and every ontogenetic event. A genome coding for specific receptors corresponding to each and every possible invader would require a DNA molecule so large as to take weeks to undergo a single replication.

The lymphocytes of mammals, despite a superficial similarity in appearance, are in fact differentiated into a number of distinct populations. Differentiation is reflected in receptor molecules arrayed on their cell membranes. Receptors corresponding to each and every potential foreign substance are not coded separately in the genome. Rather, the receptor molecules are identical in overall structure, but the various subunits of which they are composed are assembled *de novo* from one of several gene sequences chosen at random (Figure 3.7). The genetic foundation of the immune system is based on the production of random genetic variation in receptor

17. Wharton, K., T. X. Johansen, and S. Artavanis-Tsakonas. 1985. *Cell* 43:567–581.
18. *Notch* is not the only case of a developmental anomaly in which the gene controlling the anomaly includes EGF sequences. The *lin-12* mutation of the nematode *Caenorhabditis*, in which the normal somatic pathway is switched from one cell type to another, has also been found to contain at least 11 copies of EGF (Greenwald, I. 1985. *Cell* 43:583–590).

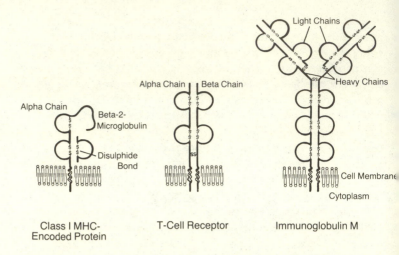

Class I MHC-
Encoded Protein

T-Cell Receptor

Immunoglobulin M

FIGURE 3.7   Cell-surface receptor molecules utilized by the immune system to amplify selected T-cell clones. The Class 1 MHC
molecules are borne on lymphocytes and act to direct macrophages to release interleukins to selected T-lymphocyte clones.
Macrophages only release interleukins, however, if the MHC receptors appear on a cell which also bears a T-cell receptor that
matches the antigen. The enormous specificity of the immune
system derives from the fact that the T-cell receptor is constructed of a constant region, a joining region, a diversity region,
and a variable region. The constant region is encoded by only one
sequence, but the joining region is encoded by one of four sequences, the diversity region by one of $>$ 10 sequences, and the
variable region by one of $>$ 100 sequences. The combinatorial diversity of T-cell receptors resulting from the construction of receptor molecules drawing on proteins encoded by different sequences is further amplified by high rates of somatic mutation in
these sequences and by imprecisions in the joining of the different regions to one another. Both the Class 1 MHC and T-cell receptor molecules display sequence homology with immunoglobu-
lins. (After Marrack and Kappler, 1986.)

molecules. The combinatorial variability arising from construction of a receptor from randomly chosen bits and pieces
leads to a large number of specific receptors.[19]

19. The capacity of the immune system to recognize an enormous diversity of antigens suggests a hypothesis for a striking trend displayed by
metazoan parasites. In the course of evolution of the parasitic habit, many
taxa have undergone a loss of characteristics found in free-living representatives. The loss of structures is frequently interpreted as "their no

Variant cell lineages, bearing their unique receptor arrangements, interact with one another systematically to amplify and suppress cell populations inhabiting the bloodstream of mammals. One compartment of this complex cascade of interactions acts to remove cells infected by foreigners. Three players are paramount in this response: the macrophage, the cytotoxic T-lymphocyte, and the helper T-lymphoctye (Figure 3.8). When an amoeboid macrophage cell engulfs a foreign object, that is, an antigen, it displays on its cell surface partially digested proteins derived from that antigen. Antigen proteins, when presented by a macrophage, can be recognized by T-lymphocyte cells. Not just any T-cell is competent to recognize the antigen. Recognition occurs only by those few genetic variants which, entirely by chance, display receptor molecules capable of simultaneously recognizing the antigen and the macrophage presenting it. Those few variants capable of binding to the macrophage are said to be selected.

Macrophages bound to the selected lymphocytes release a class of activating molecules, interleukins, which, in concert with other mitogenic interleukins, stimulate the T-cells to proliferate. Clonal selection occurs: among innumerable variants, only those clones of T-cells corresponding to the antigen presented are induced to proliferate disproportionately. The macrophage selects both lymphocyte clones by selectively activating those cells bearing receptors matching the antigen. The activated helper T-cell, under guidance from the macrophage, is induced to display receptors for mitogens that it itself produces. In addition to stimulating its own growth, the helper T-cell "helps" the selected cytotoxic T-cell by selectively releasing further mitogens to cytotoxic

---

longer being needed"; a parasite bearing traits unnecessary for parasitism is indeed burdened with the cost of construction of these features. Perhaps more important than the cost of construction, however, is the considerable antigenic diversity such structures present to the host immune system. By reducing to a minimum all structural complexity but that absolutely required for the parasitic life cycle, the parasite may reduce enormously the antigenic profile that it presents to the most immune system. If this notion has merit, then the loss of such features should be less dramatic in parasites which utilize hosts lacking sensitive recognition systems compared with parasites of hosts with elaborate devices of historecognition. A survey of the morphologic complexity of parasites with complex life cycles, in which, say, a mollusc is the first host and a vertebrate the second, may shed light on this question.

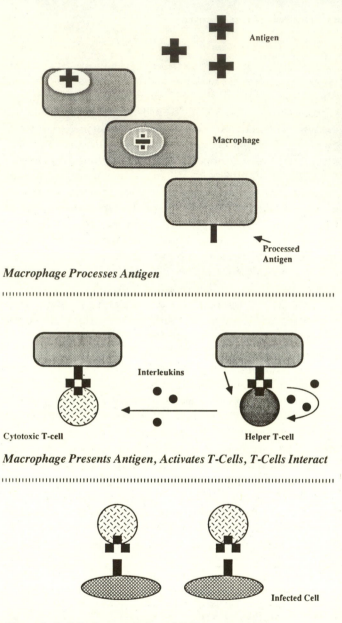

Macrophage Processes Antigen

Macrophage Presents Antigen, Activates T-Cells, T-Cells Interact

Cytotoxic T-Cells Remove Infected Cells

FIGURE 3.8   A schematic representation of a portion of the immune-regulatory network, illustrating the activation of T-lymphocytes for removal of infected cells. The foreign antigen (crosses) is engulfed by the amoeboid macrophage cells, which display partially digested protein from the antigen on their cell surfaces. T-lymphocyte clones whose receptors match the antigen

T-cells which match their receptor arrangements. The cytotoxic T-cell, amplified by mitogens from helper T-cells, is
the effector cell. It binds with infected cells and effects their
demise.

Each cell of the immune response behaves in its own self-
interest. Their *participation in somatic function occurs only because the receptors that ensure delivery of growth-enhancing mitogens
also compel somatic function*. The cytotoxic T-cell recognizes its
target with the same receptor arrangement used by the macrophage to activate that cell lineage. It is compelled to attack
the infected cell by the same receptor required for it to obtain mitogens from helper cells. Similarly, the helper T-cell
"helps" the cytotoxic T-cells because the receptor that com-
pels it to "help" is the same receptor that ensures the proliferation of receptors for its own growth factors.[20] The various
lymphocyte cell lineages cannot further their own replication without performing somatic duties. The immune system works by exploiting the inherent propensity of cells to
further their own rate of replication. Competing lineages co-
operate to produce somatic function because failure to do so
would limit their own rates of replication.[21] A central theme

20. The helper T-cell's effect on cytotoxic T-cells is not its only role in the
immune system. Helper T-cells are also crucial in the amplification and
differentiation of B-lymphocytes. B-cells are selected by macrophages and
helper T-cells and, under the influence of helper T-cells, differentiate into
plasma cells. Plasma cells release antibody which binds free antigen (as
opposed to antigen-infected cells removed by the cytotoxic T-cell). The
complex of interactions in the macrophage/helper T-cell/B-cell network
are similar to, indeed homologous with, those seen in the cytotoxic compartment: each cell participates in somatic function because the receptor
that mediates that function is the same receptor that ensures its source of
mitogens.
21. The network of interactions in the immune response is similar to that
seen in many interspecific mutualistic interactions between species. In the
immune system, the cytotoxic T-cell and the helper T-cell have a negative

---

FIGURE 3.8 (*cont.*)

are activated. The helper T-lymphocyte delivers interleukins to
the cytotoxic T-cells (dots), as well as acting to direct B-lymphocytes to produce antibodies (not shown). The cytotoxic T-cell
clone, selected by macrophages and helper T-cells, binds to infected cells and effects their demise. Note that the molecules
which allow the cytotoxic T-lymphocyte to obtain interleukins
from the helper T-cell are the same molecules which allow the
cell to recognize the antigen.

emerges from the complex network of reciprocal interactions of the immune system: a complex epigenetic system of metazoans arises as a consequence of variant cell lineages interacting with one another to enhance their rate of replication.[22]

As exemplified by the *Notch* mutation in *Drosophila*, aberrations in function occur as a consequence of cell lineages which pursue their own replicatory advantage to the detriment of the individual harboring them. That the same is true for the immune system is underscored by the occurrence of autoimmune disorders.[23] Variant T-cell lineages which

---

effect upon one another as both seek interleukins from the same source. This situation is formally analogous to that found in acacia trees, which produce extrafloral nectaries which draw both ants and herbivorous insects. The ants and the insects compete (with the result that the acacia tree is protected, to a considerable degree, from the insects). The interactions in the ant-acacia mutualism are of the same type as those in the immune system: two components repress each other's growth, while a third has a positive effect on the growth of both components. The formal similarity between these systems underscores the utility of viewing cellular interactions in an ecological context.

22. The process of producing random genetic variation in receptor molecules has likely fueled the further evolution of this remarkable regulatory circuit. Any variant which is capable of exploiting the pool of interleukins to its own benefit, but in competition with effector T-cells, will compete with effector cells and act to suppress them. Likewise, variants with similar effects on suppressor cells will act as contrasuppressor cells. Both cell types, suppressors and contrasuppressors, are known to be active components of the normal immune response. Indeed, this perspective is favored by the frequent observation that suppressor cell populations are often heterogeneous. Furthermore, selection of variants which produce modifying influences on already selected populations must act to fuel the evolution of devices to control interactions between populations. This has apparently occurred, as the immunoregulatory network is characterized by an increasingly lengthy list of highly specific mitogens which act to further select among the amplified (i.e., selected) T-cell populations.

23. Variants which actively attack self tissues are the most extreme manifestation of cell lineages which act to the detriment of the individual. More benign variants can also arise. The delivery of interleukins from macrophages to T-cells offers an obvious opportunity for variants which are capable of recognizing the macrophage, but which fail to perform an effector function. When such variants are selected, they gain access to helper T-cell derived mitogens, without the corresponding cost of participating in somatic function. Indeed, such variants are common. Response to a given antigen often yields cytotoxic T-cell clones which fail to recognize the antigen. This observation is particularly telling: if development (in this example, the development of an immune system) arose as anything but a consequence of interactions between cell lineages in the course of ontogeny, such functionless variants would hardly be expected.

inappropriately bear receptors for healthy self tissues are inexorably favored at the level of the cell lineage, but have devastating effects on the viability of the individual harboring them. Normal function is associated with synergisms between the interests of the individual and those of the cell; aberrations in function are a consequence of conflicts between the two units of selection.

The molecular underpinnings of metazoan development are destined to preoccupy biologists for the remainder of this century. The examples of the immune system and the *Notch* mutation represent only the beginning salvos in what promises to be a continuing barrage of new information. These cases provide a telling contrast. In the immune system, normal function occurs when variant cell lineages, in seeking their own replicatory advantage, serve somatic function. In *Notch*, normal function is disrupted when a variant lineage, in seeking its own replicatory advantage, fails to serve the individual in which it occurs. The parsimonious interpretation is that these vanguard examples are not aberrant, but typical. They exemplify the general explanation advocated here: metazoan development represents the fixation of variant programs in which a cell lineage in pursuing its own replication incidentally favors the individual in which it arose.

*

Owing to an unusual geological event, a snapshot of the Middle Cambrian sea (530 m.y.b.p.) was preserved in the Burgess Shale formation in northwest Canada.[24] In its fossils, metazoan life is seen blossoming. The Burgess fauna displays, in addition to several animals clearly assignable to extant phyla, several bizarre creatures which simply cannot be regarded as anything akin to organisms living on earth today. For half a billion years following the Cambrian, all metazoan life can, with some minor squabbling, be assigned to some thirty-odd metazoan phyla; yet 50 million years before the relatively well-preserved shelly deposits of the Cambrian appeared, there were no fewer than eighteen organisms which are distinct by any criterion (Figure 3.9).

Yet, the immune system, when confronted with a unique antigen, often selects functionless T-cell populations.

24. Whittington, H. B. *The Burgess Shale*. New Haven, Conn.: Yale University Press, 1985.

FIGURE 3.9

Hallucigenia sparsa, an organism from the Burgess Shale lacking affinities to any other known metazoan taxa. (After Conway Morris, 1977.)

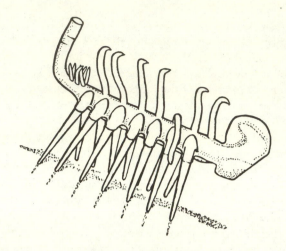

Early metazoan life was marked by a period of chaotic experimentation followed by the long, relatively conservative, march to the Recent. A precise phylogenetic reconstruction of each interaction between variant cell lineages, with a corresponding precise articulation of the various benefits to the individual and to the cell lineage which led to the establishment of modern bauplans is quite impossible at present. The fossil record simply does not provide adequate ontogenetic resolution, and surely we can expect little more to be revealed of the processes spawning the basic bauplans from the chaos of Precambrian seas. While we may hope that the newly accessible "fossil record" writ in the genome may, eventually, provide such a reconstruction, precise reconstruction is not necessary to explore further the hypothesis in question. If metazoan development indeed came about through competition between variant cell lineages arising during ontogeny, then the legacy of such interactions remains in the epigenetic programs of extant metazoans. Generations of experimental embryologists have established epigenetic mechanisms common to all metazoans—those of induction, competence, and, ultimately, organogenesis itself. We need only ask whether these fundamental epigenetic processes are events which act to allow the cell lineages engaging in them to increase their own capacity for replication.

A variant cell lineage can increase its replication rate relative to its neighbors by the simple device of inducing neighboring lineages to adopt somatic duties. Induction allows a cell lineage to increase its own replication without any loss to the organism harboring that lineage of a critical somatic function. The inducing lineage is left free to proliferate and perhaps, in so doing, to explore new functional domains. Indeed, at the end of the period of maternal control of development, inductive influences direct development. For example, the micromeres of early cleavage divisions in the echinoderm embryo introgress at gastrulation, a process followed shortly thereafter by the formulation of spicules within the blastocoele, mesenchyme surrounding the spicules, and the invagination of the blastular wall to form the archenteron. Transplantation of micromeres to unusual positions in the embryo generates spicule formation in unusual locations, mesenchyme surrounding them, and a secondary archenteron rising through the spicular-mesencyhmal mass (Figure 3.10). Micromeres hence induce the formation of germ layers. With the development of the mesoderm comes the opening of the coelom, which permits, at the level of the cell lineage, increased internal space for further replication, and at the level of the individual, all the attendant evolutionary opportunities for hydraulic design that possession of a coelom invites.

The process of induction is not limited to the establishment of germ layers, but continues throughout development.[25] For example, the invaginating endoderm of the amphibian gastrula, upon contacting the ectodermal roof in the interior of the blastula, induces the ectoderm to form the neural tube and the rudiments of the vertebrate nervous system. Still later in development, an invagination of the mesoderm approaches the ectodermal wall, inducing the ectoderm to begin elaboration of the lens of the eye. The influence of one lineage upon another such that the inducing lineage enforces further somatic differentiation upon the induced tissue pervades metazoan development. While each such inductive event in development seems precisely regulated so as to produce a functional individual, the individual inductive effects serve to allow the inducing tissue to continue replication.[26]

25. Germ layers are established under maternal direction, rather than by induction, in many taxa (e.g., molluscs, annelids, and ascidians).
26. Ultimately a limit is placed on the process of reciprocal inductions by the irreversible differentiation of one tissue. A tissue may finally reach a

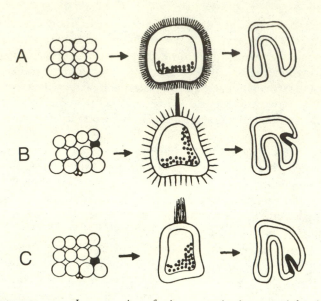

FIGURE 3.10   Introgression of micromeres in the normal development of the sea urchin (A) results in the proliferation of mesenchyme, followed by invagination to establish the endoderm. Transplantation of micromeres to unusual locations (B, C) results in the elaboration of mesenchyme at that location and the induction of a second region of invagination. (From Hörstadius, 1935.)

While induction clearly enhances the relative replication rate of the inducing tissue, the induced tissue necessarily suffers a reduction in relative rate of replication as a result of adopting the differentiated state. The induced tissue is nevertheless still free to vary, and variants which limit its susceptibility to inductive signals from competing lineages no doubt arise. Any variant within the induced lineage that renders it less susceptible to instructions from competing lineages will favor the induced lineage with a greater replication rate. Indeed, embryonic competence, the limitation of the susceptibility of a tissue to inductive signals from

state of differentiation wherein further differentiation is precluded by some ancestral constraint. To the extent that the end-point of a given inductive cascade is an irreversible differentiation event, further developmental innovation in that cell lineage in the course of evolution is limited to amplification of that end-product; no further invention of epigenetic programs from that end-point is possible.

neighboring cell lineages, is another hallmark of post-gastrulation development. Hilde Mangold, working with the
newt *Triturus*, demonstrated that transplantation of a small
segment of the dorsal lip (i.e., the site of invagination at gastrulation) from one embryo to another had a striking inductive effect (Figure 3.11).[27] Rather than simply becoming incorporated into the ectoderm of the host tissue and realizing
the developmental fate of cells it replaced, Mangold's graft
behaved radically differently. At the site of the graft a secondary embryonic axis was induced, directing first the de-

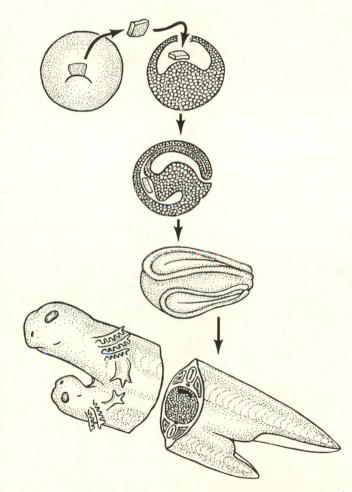

FIGURE 3.11

Hilde Mangold's ex-
periment, in which a
piece of the dorsal lip
from one newt gastrula
was implanted into the
ventral region of an-
other. The implant in-
duced a second embry-
onic axis and directed
the subsequent devel-
opment of a siamese
twin. (After Ede,
1978.)

27. Spemann, H. and H. Mangold. 1924. *Wilhelm Roux' Arch. Entwicklungsmech.* 100:599–638.

velopment of a secondary nervous system at the site of the graft and, ultimately, the elaboration of a siamese twin composed of both host and graft tissues. The cells of the host were induced to follow a differentiation pathway distinct from that which they would have originally followed. Hans Spemann, Mangold's mentor, found that this phenomenon was restricted to a brief window in ontogeny. Transplants from early embryonic donors adopted the state of the removed tissue, and late-stage transplants adopted the state that they would have realized had the transplant not been made. The competence of a tissue to be induced to form a second embryonic axis is limited to a precise period in ontogeny.

Variants limiting the susceptibility to induction limit the devices of an inducing lineage. Up to the point at which this process denies the organism a needed function, variants displaying competence should be favored at the level of the cell lineage without a corresponding negative influence on the individual. Variants limiting competence may be expected, as Hilde Mangold's experiment showed, to narrow the period of competence to the minimum necessary for organismal function. Competence, however, should be restricted solely to those tissues which have, in the evolutionary past, come into intimate contact with inducing tissues. If, as argued here, competence is an evolutionary response of induced tissues to inducing tissues, it would be expected that the only tissues which undergo selection to limit competence would be those which, in the course of normal ontogeny, come into intimate association with the inducing tissues. Indeed, it is often found that a tissue which may display quite precise competence from a given inducer can be easily induced by a tissue from portions of the embryo it never encounters during normal development.[28] Inductive signals can be highly nonspecific. For example, feathers can

28. The pervasive effects of induction led to a frenzied search for inducing substances among early experimental embryologists. Such substances were believed to hold the key to understanding development in mechanistic terms. The long and ultimately fruitless search eventually gave way to the recognition that induction was highly nonspecific. The lack of specificity has not, however, been adequately explained. The perspective of embryonic competence as an evolutionary product of spatially and temporally restricted interactions between the induced and inducing tissues provides a simple resolution of this long-standing enigma.

be induced by agar implants in epithelial tissue,[29] taste buds can be induced in odd locations by any sensory nerve,[30] and even entire limbs can be induced in the flank by nerves or nasal placodes.[31]

While embryonic competence is an important evolutionary response to induction, it is only one of several possible responses. Mangold's experiment illustrates this point dramatically. Her graft not only induced a second nervous system, but also triggered a whole sequence of subsequent inductions leading to the formation of a siamese twin. A cell lineage which had been induced to express a given somatic function by a competing lineage is not restricted in its evolutionary reaction to limiting its competence for induction. Variants arising in the induced lineage may, in turn, come to exert an inductive influence on the original inducer. Once committed to a given task, the originally induced lineage may compel its competitor to participate in the very task it initiated.

Indeed, sequential networks of induction and competence are the very processes that initiate and sustain organogenesis.[32] The development of the eye is one such case. This classical example of a highly differentiated organ of exquisite complexity comes into being by the interplay of cell lines from quite different regions of the embryo. In amphibians, the mesoderm induces the medullary plate to form the optic cup, appearing first as a protruberance growing from the developing forebrain (Figure 3.12). The optic cup eventually contacts the ectoderm and, in so doing, induces the ectoderm to form the lens vesicle which gives rise to the lens itself. Hence, the inducer of the lens itself comes into existence by induction. Only after the lens is formed, and cell division has ceased, does the cascade of reciprocal induction cease. The eye is only one such case; multiple chains of in-

29. Senegal, P. *Morphogenesis of Skin.* Cambridge University Press, 1976.
30. Zalewski, A. A. 1974. *Ann. New York Acad. Sci.* 228:344–349.
31. Balinsky, B. I. 1927. *Wilhelm Roux' Arch. Entwicklungsmech.* 110:71–88.
32. Another important developmental mechanism operating in organogenesis is cell death. Such instances can also clearly arise as a consequence of competition between cell lineages. Examples include necrosis of neurones as a result of competitive events at the termination of the axon (Hamburger, V. 1975. *J. Comp. Neurol.* 160:535–546) and the atrophy of the notochord or pronephros in mammals as a result of competition for blood supply (Saunders, J. W. 1966. *Science* 154:604–612).

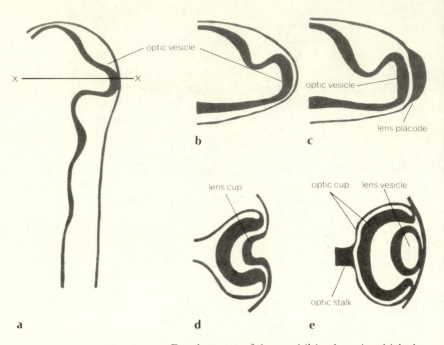

FIGURE 3.12 Development of the amphibian lens, in which the optic vesicle originates as an outpocketing of the cephalic neural tube (a), and in coming into contact with the ectoderm (b) induces the lens placode (c). The lens placode then invaginates to form the optic cup (d) and, eventually, the lens vesicle, as the optic stalk grows into contact with the developing eye (e). (From Coulombre, 1965.)

duction are typical of organogenesis (Figure 3.13). Organogenesis is the end-product of a cascade of reciprocal inductions, a coevolutionary response to repeated interactions between inducing and induced tissues.

The principal mechanisms of epigenesis involve interactions between cell lineages such that one lineage furthers its relative rate of replication as a consequence of engaging in the interaction.[33] Once a sequence of variants have arisen which, by their interplay, establish a cascade of epigenetic

33. The pervasive biomechanical constraints on ontogenetic progression are in no sense diminished by these arguments. Rather, biomechanical controls are seen as establishing the environmental milieu in which cell lineages must ultimately compete.

effects, the developing embryo must reenact the original interactions or the individual will not function. Consider, for
example, a mutant in axolotls which produces an eyeless individual.[34] The autosomal recessive mutant *eyeless* results
from a failure of the ectoderm of the *eyeless* individual to respond to mesodermal inducers, hence the failure to form an
eye. *Eyeless* mutants are not only blind, they are also sterile.
Sterility arises as a consequence of the failure of *eyeless* individuals to produce gonadotropins, through a defect in the
hypothalamus. The hypothalamic primordium resides adjacent to the eye in normal development and is induced to develop by the eye. In the *eyeless* mutant, no eye exists to direct
the ontogeny of the hypothalamus properly, and the individual is sterile. Once established, a disruption in the cascade of
inductive influences can be catastrophic. Ontogeny must
reenact the interactions which gave rise to it.[35]

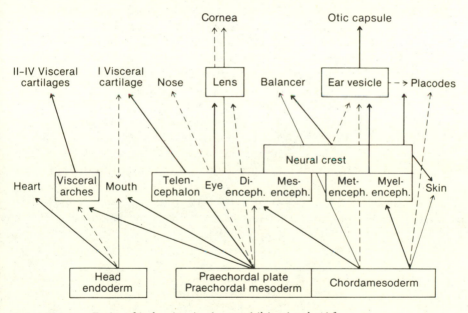

FIGURE 3.13   Paths of induction in the amphibian head. (After
Raff and Kaufman, 1983.)

34. Van Deusen, E. 1973. *Dev. Biol.* 34:135–158.
35. The processes that have given rise to patterns of metazoan development are held here to have arisen as a consequence of selection at the level
of the cell lineage. It should be emphasized that once such a trait has be-

*

An ancestral developmental progression, established by rare and fortuitous occurrences of variants favoring the individual, must surely have been endangered by the occurrence of variants which failed to behave altruistically. Detrimental variants may have one of two phenotypic effects: the detriment may cripple the cell lineage bearing it with trivial effects on the individual as a whole or, far more significantly, the replication of the variant lineage may be advantaged to the detriment of the individual as a whole. Mammalian cancers, autoimmune disorders, or the parallel cases of *Volvox* and slime molds are all instances in which a variant lineage gains replicatory advantage relative to neighboring lineages by the abandonment of somatic function to the detriment of the individual harboring them. The evolution of metazoan development required not only the exploitation of synergisms between selection at the level of the individual and at the level of the cell lineage, but also required elimination of inherent conflict between these two units of selection. Variants which abandon somatic function must have been minimized, or ancestral metazoans would not have become the stable units surrounding us today.

Just as variants which favored the individual molded epigenetic programs, and detrimental variants threatened the integrity of the individual, a third class of variants surely arose which fundamentally shaped metazoan development. Variants whose phenotypic effect preserves existing somatic functions but which act to limit the opportunity for further variants to arise, to be expressed, or to become incorporated in the germinative lineage, must have arisen at low frequency. Such a variant, by limiting the opportunity for new variants to arise, would act to fix an existing developmental program. *If such variants arose in the history of ancestral meta-*

---

come established, repetition of the same progression does not necessarily imply active selection. Rather, the pattern of differential cell growth and cell death during an established ontogenetic sequence is more precisely called "sorting." Sorting is defined as any differential birth or death process, whereas selection is defined as a category of sorting whereby differential births and deaths occur with respect to a particular heritable trait in a particular environment (Vrba, E. S. and N. Eldredge. 1984. *Paleobiology* 10:146–171; Vrba, E. S. 1984. *Syst. Zool.* 33:318–328; Vrba, E. S. and S. J. Gould. 1986. *Paleobiology* 12:217–228). In this terminology, ontogenetic events *arose* as a consequence of selection at the level of the cell lineage, but are maintained in that state by sorting.

zoans, the traits they encode would be expected to be reflected in the ontogeny of extant organisms, for the fixation of such a variant would surely act to stabilize the existing pattern of development.

Despite the bewildering diversity of metazoan development, traits which limit further production of variants are easily identified. Maternal predestination of early embryonic development is one such mechanism. A variant which, say overproduces mRNA in the germ line could have the effect of determining the embryonic program of the resulting zygote.[36] Sequential addition of similar variants would ultimately allow direction of large components of the ontogenetic process via instructions largely predetermined. Variants which establish maternal control of embryonic development act to limit the impact of any subsequent, potentially deleterious, variants. Unless the new variant arises in genes responsible for maternal direction, the variant will not be expressed phenotypically during early ontogeny and hence will not threaten proper development of the individual.[37]

Maternal control over deleterious variants is, however, severely constrained. Constraint occurs not only by the requirements of a free-living blastula, as explored earlier, but also by additional factors. Maternal predestination, by definition, requires that the egg be stocked with numerous determinants. A simple, but effective, limit on this process is given by the size of the egg itself. Total maternal control of development can occur only in small organisms or the egg would have to become impossibly large.[38] As the size of the metazoans increased, maternal predestination had to be coupled with additional controls on the production of variants.

36. This scenario is perhaps the simplest for the evolution of maternal predestination, but clearly is only relevant for sequences in which the titer of mRNA controls, directly or indirectly, the synthesis of further mRNA.

37. Maternal control over development is a particularly effective device. If maternal predetermination occurs up to the point of germ-cell determination, then developmental variants arising in the course of ontogeny cannot be expressed in that generation, nor can variants influence subsequent generations unless they affect genes encoding maternal effects. In taxa with maternal control over development through germ-line determination, fundamental evolutionary alterations in developmental mode can only arise in genes encoding maternal effects (or in genes having pleiotropic effects on maternal determination). It is hardly surprising, in this light, that such taxa are often morphologically conservative.

38. Viviparity, however, is an alternative form of maternal control over early ontogeny and can represent an escape from the egg-size constraint for some organisms (see Chapter 2, note 39).

The limitations on maternal predestination are removed by the twinned devices of secondary somatic differentiation and germ-line sequestration. Whereas somatic cells in the ancestral mode are derived from totipotent cells or from somatic cells in the same state of differentiation, the somatic cells of many taxa are capable of secondary somatic differentiation (Figure 3.14). Here somatic cells in a given state of differentiation can give rise to other somatic cells in a quite different state of differentiation. Whereas an epithelial cell of a cnidarian can only give rise to another epithelial cell, the epithelial cell of a vertebrate may, under the proper influences at the proper time, give rise to the lens of any eye or the rudiments of the nervous system which will ultimately come to innervate it. With the origin of secondary somatic differentiation, an important constraint inherent in the ancestral pattern was relaxed. In the ancestral case, a totipotent cell was required throughout ontogeny, for it alone was capable of giving rise to certain cell lines. With this ability transferred to somatic lineages, the totipotent line is no longer required to be mitotically active throughout ontogeny.

The evolution of secondary somatic differentiation is coupled with a change in the timing at which somatic cells are separated from totipotent cells. The release of the totipotent germinative lineage from the task of producing somatic tissues meant that the number of divisions made by the totipotent lineage could be reduced and, consequently, the opportunity for variants to arise became severely restricted. Taxa as divergent as the pseudocoelomate Nematoda, the protostomatous Arthropoda, and the deuterostomous Chordata employ essentially the same mechanism: their somatic cells are capable of secondary somatic differentiation and their germ lines are sequestered. Often at the early stages of development a number of cells are set aside which become the gametes. These primordial germ cells undergo a series of precise morphogenetic movements to become lodged in the presumptive gonad. They do not divide further, but rather remain quiescent until reproductive age, often frozen in the prophase of the first reduction division from the early embryo until sexual maturation. In the human, for example, the germ cells are set aside in the 56-day embryo, to remain sequestered for one to three decades.[39] Unless a variant arises

39. Just as maternal predestination can only be understood in the context

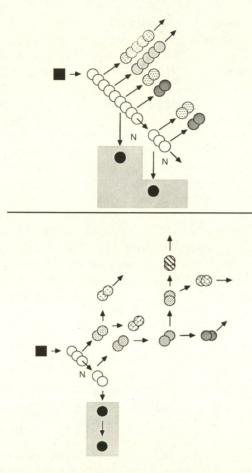

FIGURE 3.14   Schematic diagram emphasizing the distinction between development modes with and without secondary somatic differentiation. In the ancestral mode (top figure), the zygote (closed box) gives rise to totipotent cells (open circles) which give rise to somatic cells (variously filled circles) which may or may not divide further. Gametes (closed circles in shaded region) are generally derived from the totipotent lineage, which must remain mitotically active throughout ontogeny to replenish those somatic lineages incapable of division. In taxa with secondary somatic differentiation (bottom figure), the totipotent line gives rise to multipotent stem cells which elaborate somatic lineages in different states of differentiation. The totipotent lineage continues to produce the gametes, but, released from the duty of producing somatic tissues, does not continue to divide throughout ontogeny. The evolution of secondary somatic differentiation permits the subsequent evolution of germ-line sequestration.

in the first few cell divisions following fertilization it will fail to gain access to the germinative lineage.[40]

Unlike maternal predetermination, which is necessarily limited in effectiveness to small organisms, a large organism can limit the access of a variant to the heritable cell lineage by irreversibly establishing the germ line early in ontogeny. Indeed, in some groups, maternal predestination is coupled to germ-line sequestration such that the end-point of maternal control coincides precisely with terminal determination of the germ line. Metazoans, by the twin devices of maternal predestination and germ-line sequestration, have effectively closed their ontogenies to heritable intrusion arising in the course of that ontogeny (Table 1.1). A novel epigenetic program can only arise if a mutation of extraordinarily improbable precision and autonomy occurs in the germ cells themselves. With the evolution of secondary somatic differentiation, germ-line sequestration, and maternal predestination, metazoans entered a new era—one in which each generation gave rise to a new individual with minimal opportunity for variants to arise and gain access to the germ line.

It can hardly be coincidental that the simple cellular differentiating systems, like those of slime molds and *Volvox*, are demonstrably vulnerable to variants which gain access to

---

of selection at both the level of the cell lineage and of the individual (Chapter 2), so it is with germ-line sequestration. In the context of selection on the individual alone, it is difficult to comprehend why an organism might differentiate a cell lineage whose function will not be called upon for as long as 10–30 years. Allowing that germ-line sequestration limits the heritability of variants which fail to behave altruistically (that is, via consideration of the potential impact of events at the level of the cell lineage), the matter is considerably clarified.

40. Forms displaying germ-line sequestration have an important mechanism by which to "fine-tune" the extent of heritable variation allowed each generation, without such variation affecting the ontogeny of the individual in which it occurs. By allowing greater numbers of cell divisions in the germ line than are ultimately necessary for reproduction, the extent of heritable variation can be increased. In human females, for example, several million cells migrate to the presumptive gonad, while only 300 or so actually mature. The "overproduction" of primordial germ cells may represent a device to regulate heritable variation.

In this same context, male primordial germ cells undergo a greater number of divisions than do those of females. Unless the basal mutation rate is lower in males, males will contribute a greater amount of genetic variation to subsequent generations. This consideration is lacking from discussion of the "cost of males" and merits closer attention.

their reproductive tissue by failing to make somatic tissue, and that those fundamental modes of metazoan development surviving the chaotic era of experimentation of the Precambrian oceans are likewise endowed with traits which limit the production, proliferation, and access of variants to the germ line. The clear implication is that evolution of cellular differentiation fueled the evolution of controls over variants which fail to behave altruistically. The mechanisms which metazoans employ to limit the heritability of variants which fail to contribute to somatic functions are blind to the traits which a variant might express. *Potentially beneficial variants are as limited as are potentially detrimental ones*. The very features which subjugated competition among cell lines for access to the germ line and stabilized newly acquired epigenetic programs also limited the potential for the production of novelty. Once these devices are in place, so few opportunities are allowed for variation to occur that the idea that a single variation could yield an entirely new design, a new bauplan, becomes an operational impossibility. Individuality has evolved.

IV

With the evolution of individuality, stringent new boundary conditions were placed upon further modification of existing bauplans. Evolutionary innovation could occur only through the extraordinarily rare, random mutation arising in earliest ontogeny and producing a precise, largely autonomous, and beneficial phenotypic effect on an already complex functioning unit. Nevertheless, the existing bauplans have ramified into the several distinct class-level taxa and myriad less dramatic innovations within each clade. How this occurred remains largely a mystery. As reflected in the fossil record and revealed in the comparative biology of extant organisms, these are the innovations that have been the mainstay of evolutionary study. With the infusion of new comparative data from molecular studies, they will surely continue to be.

To the naturalists of the late nineteenth century, however, only the database of comparative embryology was easily accessible. Indeed, evolution by common descent owed its rapid acceptance, in no small degree, to the theory's capacity to anticipate parallels in ontogenetic pattern found within and between taxa. While Darwin's assertion that "community in embryonic structure reveals community of descent" rang true, he proffered no explicit mechanism underlying an association between ontogeny and phylogeny.[41] Ernst Haeckel, however, felt no such restraint; he was confident that phylogenetic progression arose as the terminal addition of new structures atop the ontogeny of their ancestors.[42]

> Ontogeny is a recapitulation of Phylogeny; or, somewhat more explicitly, that the series of forms through which the individual organism passes during its progress from the egg cell to its fully developed state, is a brief, compressed reproduction of the long series of forms through which the animal ancestors of that organism . . . have passed.[43]

With the rise of experimental embryology, Haeckel's biogenic law was first challenged and, with the rediscovery of Mendelian principles, ultimately abandoned. Its abandonment in no sense diminished faith that embryological detail held phylogenetic information. Comparative embryology continued, as it does today, to form the basis for phylogenetic reconstruction within and between basic bauplans. Recapitulation is simply one mode of phylogenetic impression upon ontogeny. Ontogenetic events may also be intercalated between existing stages in ontogeny or lost altogether in the course of phylogeny. While most evolutionists concur with

41. Darwin, C. *The Origin of Species*. London: John Murray, 1859:449.
42. Haeckel's biogenic law dominated the intellectual landscape of the late nineteenth century and with its abandonment came the ridicule (e.g., Garstang's poetry) that frequently follows a scientific notion proved incorrect. It would, however, be unfair to assume that Haeckel had no basis for his fervent belief in recapitulation as the foundation of evolutionary change. Haeckel's biogenic law found a comfortable companion in the Lamarckian scheme of heritability of acquired characteristics (see also S. J. Gould, *op. cit.*). The juvenile stages of an organism are transitory; only the adult stage persists long enough to face a repeated environmental demand. Hence only the adult may evolve a novel morphological device to confront a persistent challenge. Innovations, therefore, would necessarily appear in descendants at the end of their ontogeny. Haeckel's confidence in the supremacy of evolution by recapitulation had a clear mechanistic basis—a Lamarckian mode of evolutionary change necessarily yields ontogenies which recapitulate their phylogenies.
43. Haeckel, E. *Evolution of Man: A Popular Exposition of the Principal Points of Human Ontogeny and Phylogeny*. New York: D. Appleton and Co., 1879:6.

Rensch that "by far, most phylogenetic changes in form arise by heterochrony,"[44] the various modes of heterochrony degenerated into a burdensome morass of specific terminology mastered by few and frustrating to all. In the face of the increasingly experimental tenor of the early twentieth century, few taught the principles and, within a generation, few remained with a sufficiently broad command of embryological detail so as to be competent to teach them. Phylogenetic speculation fell from the pinnacle of evolutionary study that it so firmly commanded in the late nineteenth century to become the limited province of professional systematists.

Stephen Jay Gould, reflecting the pulse of modern evolutionary biology, has resurrected the problem and proposed a reclassification of phylogenetic impressions upon ontogeny.[45] His simplified taxonomy, a reduction of de Beer's original system, recognizes four classes of heterochrony: one for each of the possible relative changes in the timing of somatic versus gonadal development (Table 3.1). The revised system greatly clarifies the issue and provides a powerful guide to legitimate phylogenetic inference from embryological data. Gould's classification, however, leaves open the question of why major evolutionary change can be so neatly classified in this manner.

Why should all evolutionary reshuffling of developmental traits be reducible to these four modes—in particular to changes in the relative maturation of the gonads and the soma? These four modes should be paramount only if they reflect the mechanisms by which new additions are differ-

TABLE 3.1   Categories of Heterochrony[1]

Timing of Terminal Determination

Somatic Cell Lineage	Germ Cell Lineage	Terminology	Morphological Result
Accelerated	Unchanged	Acceleration	Recapitulation
Unchanged	Accelerated	Progenesis	Paedomorphosis
Retarded	Unchanged	Neoteny	Paedomorphosis
Unchanged	Retarded	Hypermorphosis	Recapitulation

1. Modified from S. J. Gould, *Ontogeny and Phylogeny*. Cambridge, Mass.: Harvard University Press, 1977.

44. Rensch, B. In G. Heberer, ed. *Die Evolution der Organismen. Die Kausalität der Phylogenie*, Vol. 2. Stuttgart: Gustav Fischer, 1971:12.
45. Gould, S. J., *op. cit.*

entially permitted or precluded. The fact that the most successful taxonomy of phylogenetic change is calibrated relative to the maturation of the gonads is not accidental. If one replaces the column "maturation of the reproductive organs" in Gould's classification with "terminal determination of primordial germ cells" (Table 3.1), the classification reflects the modes of addition of new features or preservation of ancestral features by a simple logical extension of the principles expounded earlier. Changes in the timing of germ-line determination act either as fuel for further evolutionary change by allowing more heritable divisions per generation, or as the brake on further evolutionary change by permitting fewer heritable divisions per generation. Changes in the timing of appearance of a somatic feature may further act either to add or delete that lineage from the stem cell line giving rise to the gametes. Gould's scheme serves as a guide to phylogenetic impression upon ontogeny *because* it reflects the underlying mechanisms whereby new features are added and ancestral features are protected from further modification.[46]

Metazoans control the heritability of variants by the twin devices of maternal predestination and germ-line sequestration. Of these two devices, evolutionary modification of maternal control over embryonic cell fate is severely constrained by egg size. In contrast, the timing of germ-line sequestration suffers no such constraint and has been frequently altered in the course of phylogeny. Differences in the number of heritable cell divisions per generation are often of astronomical proportions. The dipteran *Drosophila* terminally commits its germ line during blastoderm formation, after only thirteen cleavage divisions. Orthopteran insects, however, determine their germ line long after this stage; an orthopteran generation is the evolutionary equivalent of many dipteran generations.[47] A human commits its germ line in the fifty-sixth day of ontogeny, an urodele amphibian only

46. If most developmental innovations arose as variants in meiosis (i.e., as variation between rather than within individuals), then why should ontogenetic shifts in phylogeny be calibrated relative to the maturation of gonadal tissues? They should not. Just as the observation that metazoans develop via epigenetic interactions is *prima facie* evidence that ontogenetic mechanisms arose in the course of ontogeny as a consequence of competition between cell lineages, the fact that heterochrony is the predominant mechanism of major evolutionary change within bauplans is *prima facie* evidence that subsequent events were shaped by the same mechanisms.
47. See Chapter 1, note 17.

after its metamorphosis to an adult—after the equivalent of hundreds of human generations. Reef corals never terminally differentiate germ cells—each cell division of the totipotent lineage is heritable. A 20,000-year-old reef coral had passed uncounted millions of fruit fly generations. By Gould's re-definition, changes in these figures are, *by definition*, hetero-chrony.

At the end of the period of maternal determination of cell fate, and before the terminal determination of the germ line, a window in ontogeny may be opened to allow heritable variation to arise and gain access to the germ. This window is opened to differing degrees in extant phyla (Table 1.1). If changes in the timing of germ-line determination have fueled metazoan evolution, then major innovations should have occurred in those taxa in which the window is now, or has been in the past, opened to variation. Further, new fea-tures intercalated in the existing ontogeny should occur at that point of the life cycle at which this window is opened. Somatic features appearing during the period of maternal control should be conserved, as should be features appearing after the determination of the germ (except as cascading ef-fects of early changes).

At any given taxonomic level, comparisons between groups can be made as to the extent of evolutionary innova-tion in juvenile versus adult features. Such distinctions have been the traditional basis for phylogenetic speculation. The extent to which such changes can be inferred from the ontog-eny of extant organisms has dominated discussion of heter-ochrony for well over a century. Four broad classes are pos-sible: both juveniles and adults have remained conservative, both juveniles and adults have diverged, juveniles have di-verged and adults remained conservative, or juveniles have remained conservative and adults diverged.

Consider a phylum-level comparison. Within the proto-stomes, the Phylum Annelida and the Phylum Mollusca share a remarkably similar larval form, the trochophore (Fig-ure 3.15). Both groups display characteristic spiral cleavage, with maternal control over the initial stages of ontogeny, and a relatively late determination of the germ line. Follow-ing the conservative, maternally directed, early phase, the morphologies of annelid and mollusc embryos quickly di-verge: the trochophore of a mollusc to produce a veliger, and the trochophore of an annelid to begin iteration of somites.

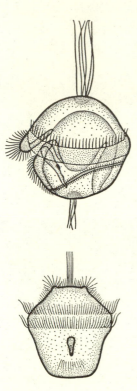

FIGURE 3.15

Trochophore larva of the annelid *Polygordius* (upper) and the mollusc *Patella* (lower). (After Wilson, 1904 and MacBride, 1914.)

Early maternal control of development restrains expression of variants and enforces conservatism in juvenile morphology, whereas late sequestration of the germ line allows the elaboration of radically different adults. The similarity of the embryos of the two phyla is underscored by the fact that J. J. Bezem and C. P. Raven were able to produce a computer simulation of the early development of the annelid *Podarke obscura* with only minor modifications of a program initially developed to simulate development of the mollusc *Limnaea stagnalis*.[48] Maternal predestination, when decoupled from germ-line sequestration, is associated with the divergence of morphologies at the end of the period of maternal control.

The annelid-mollusc pattern stands in stark contrast to phyla in which the period of maternal predestination closely coincides with germ-line sequestration—cases in which the ontogenetic window of heritability is firmly shut. In the phyla Rotifera, Mesozoa, and Chaetognatha, maternal predestination and germ-line sequestration are closely linked or even overlapping in ontogeny. Heritable intrusions in the course of ontogeny are not permitted. Each group is quite distinct and shows extreme morphological conservatism. Each taxon is characterized by a single basic design, without a corresponding series of ramifications of that design. Contrast this state to that of, say, the molluscs, whose ancestral design has given rise to forms as distinct as those seen in giant squids, whose only rivals for the dominance of the deep sea are whales; in terrestrial snails, that left the sea for the land; and in bivalves, whose design not only permits domination of the mud and sand of the sea floor, but is sufficiently flexible to sustain life in the sulfur-based food webs of hydrothermal vents. The various pseudocoelomate phyla could not give rise to such elaboration; their germ lines are closed to all variation arising in the course of ontogeny.[49]

48. 1975. *J. Theor. Biol.* 54:47–61.
49. Not only are these forms conservative within a given phylum, the phyla themselves are typically small groups. Indeed, the minor metazoan phyla, those with fewer than 1,000 species, are typically forms with extensive maternal control of ontogeny and early germ-line sequestration. The low number of species of minor metazoan groups may well be a simple consequence of their mode of development.
    If, as advocated here, periods of major developmental innovations are periods in which the germ line is open to heritable variations arising in the course of ontogeny, transitional forms would be expected to be rare. The very traits which allow these forms to change renders them unstable.

Dissimilar juveniles and similar adults are obtained by the same reasoning. If the period of maternal predestination does not extend throughout larval life and the juvenile does not have its germ line predetermined, adaptations in larval life may occur. While juvenile innovations may arise, these modifications are sure to cascade through later ontogeny. If, however, the organism undergoes a metamorphosis, larval innovations need not impinge on adult structures. If the germ line is determined shortly after metamorphosis, the larvae may be different, but adults similar. Consider a family-level example. The freshwater mollusc *Unio* is remarkable in displaying a highly modified larva (Figure 3.16), with a precocious shell modified with viciously toothed valves. The larva lurks atop the undulating mantle of its mother and clasps onto fish attracted by the mantle's movement, thereby solving the uniquely freshwater dilemma of continuous downstream loss of pelagic larvae. Unioids, despite their striking divergence in larval form, are similar to other freshwater bivalves as adults.

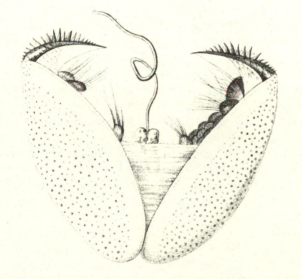

FIGURE 3.16

The glochidium larva of *Unio*, displaying sensory hairs and bear-trap valves. (From Lillie, 1895.)

Transitional forms, intermediate between modern taxa, should be rare cases in which the ontogenetic window in heritability was prematurely closed and which survive largely unmodified into the Recent as a consequence. The Onychophora, which link the annelids and the arthropods, and which display an extensive period of maternal control coupled with early terminal determination of the germ line, may be one such case.

No less important than the effect of heterochrony on the absolute rate of origin of heritable variation is the effect of temporal displacement of the terminal somatic differentiation on the rate of evolutionary modification of a given somatic tissue type. A change in the relative maturation of somatic tissues can act to change the cell lineages which may ultimately contribute to the germ line. In taxa in which the germ line is sequestered late in ontogeny, the cell lineage which gives rise to the germ has also participated in somatic function, either directly or as a stem cell. In this lineage alone a variant may prove heritable. A change in the timing of a somatic trait relative to the germ line may remove this somatic trait from further heritable intrusions. Heterochrony either removes or adds somatic or stem cell lineages to the heritable pool, with the result of either amplifying or retarding the rate of change of particular somatic tissues.

No clearer example of the impact of a change in the timing of a somatic feature can be found than that of metamerism in the annelids and arthropods. Among the annelids, the polychaeate worms display a typical spiral cleavage resulting in two large blastomeres destined to give rise to the mesoderm. These cells divide to give rise to mesoderm bands which ultimately segregate into paired somites. The somite walls then give rise to somite musculature, intersegmental septa, dorsal and lateral musculature, the major blood vessels, and the coelomoducts. Most importantly, though, in many annelids, e.g., *Tomopteris*, the primordial germ cells are derived anew in each segment and appear as histologically recognizable entities only after the major tissue types of the somite have become established. The cell lineage giving rise to the germ is the same stem-cell lineage that constructs the somite.

In some annelids, e.g., *Nephthys*, new somites are produced more-or-less continuously throughout ontogeny, with the posteriormost segment capable of asexual iteration of a new somite. Thus, variants arising in the course of ontogeny, say, modifying the form of segments or the shape of parapodia extending from them, can be both expressed and gain access to the germinative lineage in posteriormost segments. Indeed, the success of the annelids is in no small measure attributable to the evolution of specialized metameres or metameric appendages for detailed functions of locomotion, feeding, and behavior. Significantly, new metamere types appear in the posteriormost segments, the very segments in which variants arising in the course of ontogeny would be heritable (Figure 3.17).

The metameres of most extant annelids, however, no longer display a capacity for unbounded proliferation throughout ontogeny. Rather, most worms exhibit a precise, unvarying number of segments, established early in ontogeny. An important somatic feature, the number of metameres, has become fixed early in ontogeny while the gonads continue, in many species, to be determined late in ontogeny. While variants can still arise in the course of ontogeny, they cannot produce the serial repetition of that variant. Hence metameres which "discover" a new design of potential value to both the metamere and the individual are no longer capable of continuously iterating that design.[50]

While the remarkably rich tradition of annelid descriptive

50. While metameres do not proliferate throughout ontogeny in most modern annelids, this was probably not the primitive condition. Primitively, somites were likely capable of both replication (the unbounded production of new somites) and variation (the transmission of a variant design to all daughter somites of the variant). If so, metameres acted as a unit of selection—a unit above the level of the cell and below that of the individual. Just as in the case of the cell as a unit of selection, the metamere as a unit implies both synergisms and conflicts between selection on the metamere and on the individual. While the unbounded proliferation of metameres has been restricted by early embryonic determination of metamere number and type in most modern taxa (just as in the case of cells), this restriction presumably occurred only after synergisms established a variety of new metameric designs. This perspective may provide simple models for the loss of metamerism in organ systems (e.g., leeches), the stereotypic distribution of gonads within particular metameres (e.g., oligochaetes and various polychaetes), the distribution of heteromorphic metameres between species within genera, families, and orders (e.g., in polychaetes), and the occurrence of otherwise peculiar alterations of sexual and asexual phases (e.g., in polychaetes). Indeed, the diversification of the various higher protostome phyla may ultimately be interpreted in the context of metameres acting as units of selection in their ancestral stock.

In this context, another potential unit of selection above the level of the cell, but below that of the individual, is that of the polyp or zooid of colonial invertebrates. Each such unit is capable of potentially unbounded iteration and of transmissible variation to all daughter polyps derived from a variant. In this perspective, synergisms between selection at the level of the polyp and that of the individual acted to establish heteromorphic polyps (e.g., hydroids, bryozoans, and tunicates) and unusual forms of alternating generations (e.g., cnidarians), whereas conflicts between the two units fueled the present restriction of gamete production to specialized zooids and/or life stages.

FIGURE 3.17

Chaetopterus, an annelid displaying substantial differentiation in the structure and function of metameres. (From Douce, 1865.)

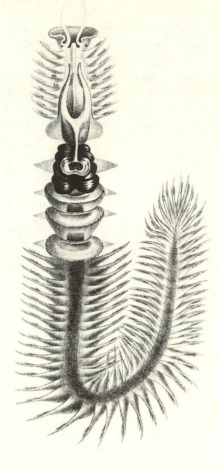

embryology invites speculative phylogeny, these organisms have not attracted the attention of investigators using molecular tools. Here we must turn to the arthropods. If evolution has indeed proceeded by the sequential addition of variant programs arising in the course of ontogeny, followed by the differential preservation of these programs either by germ-line sequestration or early terminal differentiation of the somatic fate, it should be possible to reconstruct this sequence, in a rough sense, through the appearance of atavisms. If conservatism of a particular somatic feature represents a heterochronic shift in the timing of terminal determination of that trait, then a mutant reversing early determination should reveal a design similar to that hypothesized to represent the phyletic precursor. Atavisms, as evolutionary throwbacks, provide vital clues to the manner and sequence of modifications on a given design.

Atavisms appear in a number of genetic contexts. Recently, tremendous effort has been directed to the task of mapping the phenomenology of homeotic mutations of genes affecting segmentation in *Drosophila* to their underlying molecular architecture. The most important homeotic mutations from this perspective are those which delete particular genes, for deletion of the gene controlling early differentiation should then reveal an ancestral design. Deletion of the *Ultrabithorax (Ulb)* sequence results in an iteration of the second thoracic segment and loss of the third thoracic segment, with dramatic effect. The drosopholids are pterygotes, two-winged insects, derived from four-winged insects. *Ulb*-individuals bear four wings; deletion of a single gene controlling early determination of somatic cell fate produces a pterygote fly which mimics the segmental design of more primitive apterygote insects. The apterous condition is further revealed by deletion of the *Antennapedia* gene, which produces three similar thoracic segments, entirely removing thoracic segmental differentiation. A design strikingly reminiscent of the trignathous myriapods, a totally different class of arthropods from which apterygotes are believed to be derived, occurs if the entire Bithorax gene cluster is deleted. Here the various abdominal segments are transformed into a serial repetition of thoracic segments. Finally, if both the Bithorax and Antennapedia gene clusters are deleted, all but the anteriormost segments are transformed into thoracic repeats. This is the metameric plan of onychophorans, a small phylum classically regarded as the "missing link" between the arthropods and the annelids (Figure 3.18).[51] As Raup and Kaufman stress,

> . . . the atavisms presented are not actual. What has been changed by deleting genes is the pattern of segment identity. The segments produced are still undeniably drosophilid in character. This indicates that what these loci represent are genes involved in the spec-

51. A final atavism might be imagined in this sequence whereby the various segments reacquired gonads. Indeed, Richard Goldschmidt (1923. *Archiv. Mikr. Anat.* 98:292–313) found such an atavism in gypsy moths in which reproductive organs were repeated in each abdominal segment. In fact, in the primitive insects *Iapyx* and *Machilis*, the female reproductive organs are iterated in normal development.

FIGURE 3.18

Segment design in an (A) annelid, (B) ony-chophoran, (C) myria-pod, (D) apterygote in-sect, and (E) pterygote insect. Superimposed on this gross phylog-eny are the segment or-ganizations produced by deletion of home-otic genes in the ptery-gote fly *Drosophila*. (After Raff and Kauf-man, 1983.)

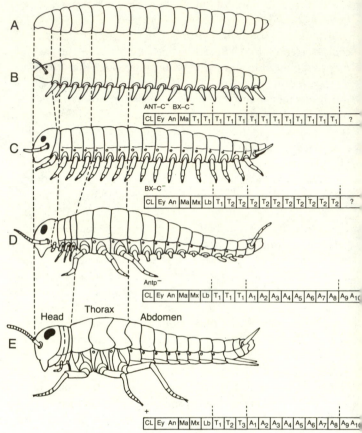

ification of pattern—the control of ontogeny *per se*. . . . The structural genes under their control must have evolved concomitantly with, but at the same time in some sense, separately from the controlling homeotic loci.[52]

The sense in which structural genes have evolved separately from the control of segmentation is the theme developed here: structural specializations arose as heritable modifica-tions of iterating segments; shifts in the timing of determi-nation of these segments, controlled by homeotic genes, sta-bilized this pattern.

52. Raup, R. A. and T. C. Kaufman. *Embryos, Genes, and Evolution*. New York: Macmillan, 1983:259.

Variants arising in clonal cell lineages within a multicellular organism are held here to have both established the epigenetic programs of metazoans by their pursuit of their own replicatory advantage, and to have stabilized those very programs by protecting the individual from variants which failed to behave altruistically. Individuality evolved and thereby restrained further heritable variation arising in the course of ontogeny. Yet, the processes which gave rise to the individual are not completely fixed. Changes in the timing of maturation of the germ and of those somatic lineages with access to the heritable lineages have occurred. Such changes control the rate at which variation arising in the course of ontogeny will be inherited. *Heterochrony is a predominant mechanism of evolutionary change within established body plans because heterochronic change resets the boundary conditions on further change and allows the same processes which ultimately spawned those body plans to act to modify them.* Some such variations leave a phylogenetic record on ontogeny, as von Baer, Darwin, Haeckel, Garstang, de Beer, and Gould have sequentially explored. The most successful taxonomy of this record is one which reflects the processes which ultimately gave rise to it.

V

Nature has the unfortunate habit of rewarding the scientist who asks a misdirected question with little more than an invitation to further misdirection. The most vexing scientific problems are often left unresolved not for lack of adequate information or appropriate technology, but for a failure to specify precisely those issues that require explanation and those that do not. In evolutionary problems, assumptions as to the state of ancestral organisms are paramount, for without a precise articulation of the ancestral characters we can never be clear on the degree to which subsequent evolutionary events are the logical extensions of characteristics already fully realized by ancestral forms. Indeed, the history of evolutionary biology, as explored at length in Ernst Mayr's *The Growth of Biological Thought*, is often traceable to scientists' propensity to

> rarely articulate—if they think about it at all—what truths or concepts they accept without question and what others they totally reject . . . an understanding of

these silent assumptions may be necessary in order to answer previously puzzling questions.[53]

A "silent assumption" was born in the tacit acceptance of Weismannian dogma by the framers of the Modern Synthesis. Inherent in the view of the individual as the sole unit of selection is the notion that heritable variation arising in the course of ontogeny is of no great phylogenetic significance.

While this assumption is surely justified for modern metazoans with overlapping periods of maternal direction and germ-line sequestration (e.g., *Drosophila*), there is no particular reason for assuming that this condition prevailed when the major innovations in metazoan development first evolved. Indeed, there is every reason to believe just the opposite: a primitive multicellular organism has *two* routes by which heritable variation might arise. Variants may arise at the onset of a new generation, during meiosis, producing an individual composed of genetically homogeneous cells; or variants may arise in the course of ontogeny, during mitosis, producing a chimeric individual composed of the original and variant lineages. Of the two modes, the latter must certainly have been the most common mode in ancestral forms for the simple reason that meiotic divisions occur once per generation while mitotic divisions occur innumerable times.[54]

Variants arising in the latter mode must interact with the cells which gave rise to them. These interactions have consequences not only for the fate of the cell lineages involved in the interactions, but also for the individual harboring them. A variety of phenotypic effects of variants, both on the level of the cell lineage and of the individual, are conceivable. The variant may increase its own rate of replication by channeling the growth or differentiation of neighboring cells, with the entirely fortuitous effect of benefiting the individual as a unit. If such a variant proved heritable, the history of its origin from the original lineage and its interaction with the original lineage may be replayed as an epigenetic interaction in development. Benefit to both the individual

53. *Op. cit.*, pp. 17–18.
54. The role of mixis in dispersing genetic variation from the individual in which it first arose to other individuals in the same population is often confused with the site of origin of heritable variation. Most variation *arises* in the course of ontogeny, but *disperses* through a population by sex.

and the cell lineage, however, is only one of several possible phenotypic consequences of variation arising in the course of ontogeny. A second class, whose phenotypic effect is to benefit the cell lineage at the expense of the individual, routinely occurs. This second class of variants and the mortality threat it poses may be controlled by yet a third class of variants, those whose phenotypic effect is to limit the production, expression, and/or proliferation of further variants. This third class, manifested as germ-line sequestration and maternal predestination, acts to stabilize the existing pattern of epigenesis, limit the heritability of variants, and enforce a rigid distinction between the soma and the germ. Once these devices were in place, the interests of the somatic lineage became identical to the interests of the organism harboring it. Only if that individual reproduces will ontogeny be replayed and cells of metazoans be free to explore again the consequences of further replication.

Richard Goldschmidt, one of many who have attempted to envision a mechanism by which the complex epigenetic programs of metazoan development may have evolved, concludes:

> If one tries to work out this idea in detail one soon comes to a point where it is evident that something besides the Neo-Darwinian tenets is needed to explain such processes.[55]

Given the Neo-Darwinian tenet of the individual as the sole unit of selection, the dilemma is real and unresolvable. Relax the assumption that evolution is solely attributable to selection on the level of the individual, and the major features of metazoan development appear as a logical extension of facts already firmly established and agreed to by all.

55. Goldschmidt, R. *The Material Basis of Evolution*. New Haven, Conn.: Yale University Press, 1940.

LIFE CYCLE EVOLUTION

A plausible argument could be made that evolution is the control of development by ecology. Oddly, neither area has figured importantly in evolutionary theory since Darwin. —VAN VALEN, 1973

Summary

With the evolution of individuality, selection on the vegetative stage of the life cycle occurred not only on cells, but also upon the individual. For the first time, each potential innovation in the vegetative phase passed through a new selective filter. Those innovations which benefited both the cell and the individual persisted; those which created a conflict between the differing units of selection were restricted to life cycles capable of resolving the conflict. In turn, the modes of development eventually adopted acted differentially to restrict or enhance further life cycle elaboration. Quite different limits on the diversity of possible life cycles were ordained by the very different paths which individuality followed in the three multicellular kingdoms. The ancestral characteristics that each clade brought to this fundamental evolutionary transition defined the somatic context in which cell lineages interacted, and thus determined the nature of potential synergisms and conflicts between selection on the level of the cell and that of the individual.

Animal cells can move. Therefore their integrity could be threatened by cellular variants which abandoned supportive function, and animal development became increasingly marked by early embryonic determination of cell fate. Early germ-line sequestration precluded asexuality, as its evolution required the loss of a mitotically active totipotent lineage. Secondarily derived asexual forms arose via modifications in that brief period of ontogeny between mixis and germ-line sequestration and, by effectively amplifying the period prior to germ-line determination, acted to fuel the evolution of complex life cycles among secondarily asexual forms. The potential of intraspecific fusion to lead to subsequent parasitism by one component of the chimera made possible the evolution of elaborate mechanisms of historecognition acting to limit fusion to close kin. As diploid cells are competitively superior to haploid cells within the somatic environment, in part due to a lower per capita contribution to somatic tissues resulting from their large size, diploidy came to dominate the proliferative phase of animal life cycles.

Fungi, starting from different ancestral conditions, followed a rather different progression. As they lack cells, the basis for diploid superiority in cell size was removed, and fungi retained their prim-

itively haploid condition. Yet, being coenocytic, fungi are nonetheless susceptible to the deleterious effects of variants, and they evolved highly specialized patterns of cell-cycle synchronization to limit the relative replication rates of nuclei. They also evolved specialized forms of septation and controlled cellularization to limit the free movement of nuclei within the mycelium. Fusion between individuals invited parasitism, leading to the evolution of historecognition systems. Recognition systems function either to limit fusion to close kin or to locate and exchange nuclei between unrelated mycelia. The latter became a highly controlled phenomenon, fueling the evolution of the heterokaryon as a unique form of sexuality and forming the basis for elaborate pathogenic life cycles.

Plant cells, possessing rigid walls, cannot move, and necessarily preserve an apically totipotent lineage as a consequence. By virtue of the inability of variant cells to move from their site of origin, variants can never become systemic. Plants are therefore inherently less susceptible to deleterious variants arising in the course of ontogeny. Having thus escaped the necessity of early germ-line determination, plants could preserve asexuality. Similarly, the inability of variants to migrate following fusion removed any requirement for historecognition. The restriction on cell movement also removed the inherent advantage of diploid cells over their haploid counterparts, allowing both the retention of a haploid proliferative phase and the intercalation of a diploid proliferative phase between haploid generations.

While life must cycle if evolution is to proceed, the modern synthetic theory places no inherent restrictions upon the units of organization which might alternate, nor does it provide insight into the multitudinous variations which have arisen in life cycle evolution. The evolution of development and the evolution of life cycles are necessarily interdependent. A coherent theory of ontogeny is a prerequisite for producing a theory of life cycle diversity. The beginnings made here, involving only seven different sets of traits, allow nearly a thousand conceivable life cycles. Consideration of the role of life cycles in mediating potential conflicts between units of selection, and the role of development in subsequently placing limits upon life cycle elaboration, reduces the number of possible life cycles to a mere twenty-seven, well within the range of life cycle diversity actually known to occur.

I

Evolution can occur only if life cycles. Life itself does not require reproduction—sterile individuals are no less living entities for their misfortune—but life without reproduction is surely doomed. Only if life reproduces itself can it persist indefinitely. And only via reproduction can natural selection have products on which to act. Life cycles are the *sine qua non* of evolution. Yet, as George C. Williams says, "The main work of providing a workable theoretical structure for understanding the enormous diversity of life cycles remains to be done."[1] The condition is curious; on the one hand we have a modern evolutionary theory, with which we may quibble but which we do not fundamentally doubt, and on the other hand we have no widely accepted explanation as to why the very feature that permits evolution to occur takes on the myriad forms that it does.

The diversity of modern life cycles is extraordinary: some organisms are unicellular, some multicellular, some coenocytic; some multicellular organisms display cellular differentiation, some do not; some organisms are sexual, some asexual, some both; some organisms are vegetatively haploid, some vegetatively diploid, some both; and so on in a vast combinatorical array. However, the rich variety of life cycles is somewhat deceiving. While much has proved possible in life cycle evolution, much apparently has not: there are no primitively haploid animals, there are no organisms with cellular differentiation that are not primitively sexual, there are no clonal organisms with germ-line sequestration, and so on. Of the possible life cycles, evolution has chosen only a subset. Somewhere in the sequence of evolutionary innovation, limitations on the diversity of life cycles must have arisen. An understanding of the relationship between that which is actual and that which is possible in life cycle evolution must surely lie in exploration of conflicts and synergisms between those units of selection that alternate in the life cycles of extant organisms, that is, in the conflicts and synergisms between a single-celled and multicellular state.

II

Sexuality is a seemingly ubiquitous component of the life cycle of cellular-differentiating organisms. The great frequency of sexuality, like so many other biological problems, was first perceived as an evolutionary enigma by August

1. Williams, G. C. *Sex and Evolution*. Princeton University Press, 1975:119.

Weismann.[2] Weismann attributed the ubiquity of sex to its role in creating genetic variation upon which natural selection could subsequently act. He wrote:

> Sexual reproduction will readily afford such combinations of required characters, for by its means the most diverse features are continually united in the same individual. I do not know what meaning can be attributed to sexual reproduction other than the creation of hereditary individual characters to form the material upon which natural selection may work.[3]

Weismann's prescience is remarkable, particularly so in that this interpretation was promulgated during a period when biologists were far from unanimous in attributing heritability to the contents of the cell nucleus. While numerous recent investigations have confirmed Weismann's ideas as to the efficacy of sex in distributing variation between individuals within a population, these same investigations have not established that the ubiquity of sex is directly attributable to this effect. John Maynard Smith, in reaction to existing explanations for the ubiquity of sex, laments that "one is left with the feeling that some essential feature of the situation is being overlooked."[4]

George C. Williams, whose *Sex and Evolution* initially redirected the attention of evolutionists to the problem of sexuality, likewise concludes that population (or individual) advantage is alone insufficient to account for the ubiquity of sex. Williams ultimately resorts to a historical perspective. The phylogenetic distribution of any trait may be approached in this fashion: Is the character in question shared by all organisms of a given taxonomic grade? Are exceptions, cases in which the character is lacking, clearly attributable to secondary loss of the character in question? If so, the trait is probably ancestral and the frequency of the trait is less a

---

2. Weismann was fully aware of what is now referred to as "the cost of sex." In 1889, he wrote: "As soon, however, as parthenogenesis becomes advantageous . . . it will not only be the case that colonies which produce the fewest males will gain advantage, but within the limits of the colony itself, those females will gain an advantage which produce eggs that can develop without fertilization." (Weismann, A. *Essays on Heredity and Kindred Biological Problems*. E. B. Poulton, S. Schonland, and A. E. Shipley, eds. and trans. Oxford: Clarendon Press, 1889:326.)
3. *Ibid.*, p. 281.
4. Maynard Smith, J. 1976. *J. Theor. Biol.* 63:245–258.

consequence of the trait itself than of the relative success or failure of the entire bauplan(s) displaying the trait.

The overwhelming majority of organisms which display cellular differentiation are sexual. Exclusively asexual, cellular-differentiating organisms are, however, known. The mode of reproduction in such forms is, without question, derived from an ancestral sexual state. Asexuality, as the sole mode of reproduction, occurs only as an evolutionary modification of the sexual process in these taxa. The ubiquity of sex invites a simple historical interpretation: sex is common because sex is ancestral in all organisms with cellular differentiation. The parsimonious explanation is that sex is widespread because cellular differentiation is widespread.[5]

The fact that cellular-differentiating taxa are primitively sexual may reflect an entirely fortuitous correlation of the two traits. The far more interesting alternative, though, is the possibility that the association between the two traits is causal, that it reflects a fundamental constraint on the evolution of cellular differentiation. Sex may have been a necessary precondition for the evolution of cellular differentiation. If this latter alternative is correct, the absence of primitively asexual cellular-differentiating organisms reflects a severe and fundamental restriction on the diversity of "permissible" life cycles.

Simple cellular-differentiating systems are demonstrably vulnerable to variants arising in the course of ontogeny which abandon somatic functions, parasitize the somatic contributions of neighboring cells, and relentlessly increase in frequency by virtue of their failure to support somatic function. The occurrence of such "somatic cell parasites" is

---

5. Graham Bell (*The Masterpiece of Nature*. University of California Press, 1982:90) has voiced, in an unusually explicit form, the reluctance of many evolutionists to accept a historical explanation for the ubiquity of sex, writing: "A more fundamental reason for objecting to the historical hypothesis is that it violates the axiom of perfection. It is well-known that theoretical population geneticists prefer equilibrium theories and tend to discount the role of history." Bell despairs that "the only prediction generated by the historical hypothesis is that the correlates of parthenogenesis should be taxonomic rather than ecological," and concludes, "I shall continue to believe that history is bunk." This is quite remarkable, as his critical review of metazoan parthenogenesis clearly shows a systematic bias to parthenogenesis. While one cannot help but share Bell's frustration that historical processes are frequently impenetrable, we can hardly use their intractability as the basis for contending that such processes do not exist.

of great importance to our attempt to reconstruct the evolution of cellular differentiation. A multicellular organism is populated by cells all competent to reproduce. Consider the fate of a variant that first displays cellular differentiation arising within such an organism. A cell that produces solely somatic descendants will, of course, fail to reproduce itself. Its descendants cannot give rise to a new individual, by definition. If, however, the variant were to retain totipotency in some percentage of its subsequent divisions—say it produced a somatic cell every third division—it would nonetheless produce far fewer propagules than those neighboring cells which expended no effort in producing somatic tissue. Variants which display cellular differentiation in a multicellular organism are at a severe competitive disadvantage in the somatic environment. Variants which display cellular differentiation cannot increase in relative frequency when rare; they will always be competitively inferior to totipotent cells in the somatic environment in which they arise.

Consider an asexual multicellular organism that "discovers" cellular differentiation in one of its lines. A low percentage of the descendants of the original differentiating cell lineage might survive the somatic filter to initiate a new differentiating clone. However, back-mutations which reverse the state of differentiation will—sooner or later—arise and increase in relative frequency within the somatic environment. Such reversions are inevitable: an organism must periodically free itself of them if it is to persist. However, asexual organisms have no mechanism for purging themselves of mutations, short of the extinction of the clone itself.[6] As such, asexual organisms which evolve cellular differentiation will always eventually revert to the multicellular state by virtue of the origin of variants which abandon differentiation

6. This is Muller's ratchet (1964. *Mut. Res.* 1:2–9). Muller pointed out that "an asexual population incorporates a kind of ratchet mechanism, such that it can never get to contain, in any of its lines, a load of mutation smaller than that already existing in its at present least-loaded lines." Muller's argument makes the valuable point that sexuality can purge a system of mutants. The ratchet (see also John Maynard Smith, *The Evolution of Sex*. Cambridge University Press, 1978:34–36) was formulated to address the regulation of mutation load in sexual versus asexual organisms and, like all modern population dynamic theories, is directed solely at the level of the individual. The point Muller raises nevertheless takes on special significance when one considers its implications for the impact of selection within the somatic environment.

and then competitively exclude co-occurring differentiating cell lineages within the somatic environment.

A sexual multicellular organism has one distinct advantage over an asexual one in this context. Sexuality provides for the periodic birth of new individuals which are free of back-mutations reversing the differentiated state. While an asexual taxon can never free itself from the impact of the eventual appearance of somatic cell parasites, the sexual organism may. Sexuality will produce new individuals whose offspring are not chimeric. Competition within the somatic environment, which must inexorably oppose the evolution of cellular differentiation, need not occur. Sexual reproduction allows the routine reinitiation of an individual within which all cells contribute to somatic duties, hence allowing a cellular-differentiating variant, which could not survive in a chimeric somatic environment, to succeed.

The role of competition within the somatic environment in maintaining cellular differentiation can be readily demonstrated. Sexual organisms, if sexuality is artificially restrained, are no less vulnerable to the occurrence of back-mutations and the relentless competitive advantage of such parasites over normal cells than are asexual taxa. The initial advantages of sexuality may be rapidly swamped under these circumstances. Indeed, the occurrence and subsequent proliferation of somatic cell parasites are the bane of laboratories which maintain simple cellular-differentiating organisms in continuous asexual culture. Such cultures inevitably "go bad." In slime molds, for example, the progressive deterioration of cultures is often manifested as a gradual decrease in the height of fruiting bodies; this is a consequence of the relative decrease in the ratio of somatic stalk cells to reproductive spore cells.[7] Slime molds, as alluded to earlier, are par-

---

7. Slime mold cultures are routinely propagated either by flooding the petri dish to collect spores or by running a wire loop across the surface of fruiting bodies. In my experience, cultures collected in the former fashion "go bad" via a gradual decrease in stalk height, whereas cultures collected in the latter fashion "go bad" via a progressive increase in stalk height. Collecting with a wire loop unwittingly imposes strong selection on the individual against somatic cell variants, by favoring only those strains which produce large fruiting structures. Collection by washing imposes no such selection against variants, and hence somatic cell parasites increase in frequency. The routine laboratory maintenance of cultures maintained clonally may provide a striking demonstration of the conflict between selection at the level of the cell lineage and of the individual.

ticularly susceptible to such somatic cell parasites, which abandon the somatic duty of producing a stalk and parasitize the stalk-producing tendencies of others. A similar situation occurs in fungi maintained in asexual culture. For example, Miller found that continued asexual propagation of *Fusarium* led to frequent development of so-called "patch-mutants," regions of the mycelium in which normal hyphal growth was diminished and an excess of asexual propagules formed, whereas cultures newly established from the soil produced quite normal mycelia.[8] Through the origination and proliferation of back-mutations, the advantages of sexuality for the evolution of cellular differentiation can be removed. If, however, sexual maturity is reached before there is substantial proliferation of back-mutations to the multicellular state, the cellular-differentiated state will persist because sexuality will periodically purge the system of competitively dominant somatic cell parasites.[9]

The association between sex and cellular differentiation is causal; *sex was a necessary precondition for the evolution of cellular differentiation.* The evolutionary significance of sexuality lies not only in its potential advantages to the individual and the population, but also in its impact upon selection at the level of the cell lineage.[10] By allowing the individual to override

8. Miller, J. J. 1945, 1946a,b. *Can. J. Res.* C23:16–43, C24:188–212, 213–223.

9. W. D. Hamilton (1980. *Oikos* 35:282–290) has argued that the ubiquity of sex is attributable to the action of parasites. In his view, sexuality ensures that the life cycle is reinitiated free of pathogens. While his arguments are developed with an explicit view toward the interspecific interactions between parasites and their hosts, the notions presented here are parallel. I argue that the evolution of cellular differentiation hinges on the ability of an organism to free itself of variants which act as parasites on somatic cells, hence a strictly intraspecific—indeed, intraorganismal—focus. The argument proposed here is merely an extension of Hamilton's (and Muller's) ideas to encompass parasites acting in a somatic environment.

10. This interpretation is in no sense mutually exclusive to those emphasizing sexuality as the principal device for spreading genetic variation throughout a population. Quite the contrary, while sexuality permitted the initial adoption of the differentiated case, it hardly eliminated the origin of variation in the course of ontogeny. As prior discussion has held, the origin of such variants established the complex epigenetic programs by which metazoans develop. Sex was a preadaptation for the evolution of development not only because it allowed periodic removal of somatic cell parasites, but also because it rapidly spread those rare variant programs, arising in the course of ontogeny of a single individual, throughout a population.

the selective advantage of the cell lineage, sex allowed the evolution of a trait that is disadvantageous in the somatic environment. Not just any life cycle is possible.

III

In the life cycle of all modern cellular differentiating taxa, sex is followed by the development of a multicellular individual from a unicellular state. Development, as explored at length earlier for metazoans, has followed paths influenced to a considerable degree by the competing interests of the cell lineage and the individual. However, metazoan development is, after all, unique to metazoans. Cellular differentiation has evolved in several other clades. If an attempt is ultimately to be made to explore the diversity of life cycles in different kingdoms, we must first account for the principal differences in development displayed by other groups.

Each independent phylogenetic experiment in development began with its own peculiar set of boundary conditions; the ancestors of each clade brought to that lineage their own distinct history of prior specialization. Despite fundamental differences in initial conditions, each clade which gave rise to cellular differentiation was faced with an identical selective challenge: variants will surely arise within the somatic environment to threaten the integrity of the individual as a whole. In the face of this novel challenge the primitive plants, animals, and fungi were endowed with differing repertoires of prior innovations. These prior innovations defined the somatic context in which variants arose and competed. The very different developmental progressions explored by the three kingdoms must, to some considerable extent, reflect the opportunities that these primitive traits presented for exploitation and the constraints these same traits placed on further evolutionary innovation.

Of the multiple attempts at cellular differentiation, kingdoms arose from only three groups. The Kingdom Fungi, the Kingdom Animalia, and the Kingdom Plantae differ fundamentally in aspects of both their basic subcellular and cellular architecture, in traits that are surely ancient. The Fungi are rigid-walled and coenocytic; hyphae are primitively acellular with nuclei free-living within a common cytoplasm. The Plantae are also rigid-walled, but, unlike fungi, are uniformly cellularized. The Animalia, in contrast to plants and fungi, are characterized by non-rigid cell membranes. Three taxa which evolved cellular differentiation to become the modern kingdoms differed fundamentally with respect to two traits: their state of cellularization and the rigidity of their cell membranes (Table 4.1).

TABLE 4.1

Kingdom	Cellularized	Rigid Cell Walls
Fungi	No	Yes
Plantae	Yes	Yes
Animalia	Yes	No

The proliferative phases of the life cycles of plants and animals differ markedly. Most notably, the zygotes of animals give rise to differentiated lines in the embryo which are destined for functions which appear only much later in ontogeny. The metazoan zygote develops via a pattern of early establishment of the major tissue and organ systems of the adult; post-embryonic development is largely a matter of growth and relative enlargement of a body plan established in the early stages. The concentration of morphogenetic activity at early stages is made possible by the ability of cells to move relative to one another, adopting positions which, upon enlargement, will give rise to the appropriate form of the adult. The non-rigid cell walls of metazoans place no inherent constraint on cell movement and thus allow such a developmental modality to arise. While the capacity of animal cells to move rendered the metazoans vulnerable to variants which moved from a purely somatic position to a position destined for the germ line, this same capacity provided the opportunity for subsequent evolutionary tinkering, in the form of early somatic and germ-line differentiation, to render harmless the very threat it first made possible.

Animals, then, have restricted the heritability of variants by decoupling differentiation in space and time. Seemingly uniform embryos divide, differentiate into one of several epigenetic programs in earliest ontogeny, and eventually elaborate the adult body plan through complex patterns in relative growth. Contrast the early embryonic determination of metazoans with the development of plants. The rigid cell walls of plants limit the capacity for cell movement. The rigidity of cell walls effectively requires that cellular differentiation in space and time cannot be decoupled. An organism with rigid cell walls cannot begin to differentiate a structure embryonically and bring it into play at some later point in ontogeny. The constraint on movement demands that at least one cell line remain totipotent and mitotically active throughout ontogeny, because the movement of totipotent cells to the site of growth, somatic differentiation, and reproduction is impossible. Whereas an animal may be characterized by preformistic development, whereby the heritable lineage is predetermined at the earliest stages of ontogeny, a cellularized organism with rigid cell walls must develop by somatic embryogenesis; the apical lineage must retain heritability throughout ontogeny. No other option is available.

Since rigid cell walls demand a developmental program in which one cell line remains active and totipotent, will not organisms with this feature remain susceptible to exploitation by potentially parasitic cell lines? To a considerable extent this is not the case. Just as rigid walls constrain the normal cell lines, so do they constrain the variant. If cells are not free to move from one region of an organism to another, an important class of potentially detrimental effects is rendered relatively harmless. Detrimental variants may arise, but they may never become systemic. The magnitude of this advantage is readily apparent by comparing the widespread occurrence and devastating consequences of metastasis of malignant tissue in animals with the relatively benign and localized nature of plant tumors.

Cellularization with rigid walls not only demands that development be of the somatic embryogenic sort and that the threat of somatic cell parasitism be minimized, it also preserves the potential for beneficial variants to be inherited. Variant cells arising in the totipotent meristematic tissues of a plant will succeed as a function of competitive interactions with other meristematic lineages within the same plant. It is perhaps not coincidental that metazoans, whose germ lines became largely closed to heritable intrusions in the Precambrian, have not elaborated a new bauplan since that time, whereas plants, whose germ line remains open, have elaborated several new designs over the same interval.

The heritability of variants arising in the course of the ontogeny of plants is well known and is, in fact, the basis for considerable commercial activity. Meristematic variants, known as bud-sports, are the origin of many, if not most, agricultural varieties. The heritability of potentially beneficial variants in plants may be of tremendous importance as a mechanism permitting microevolutionary change during the life span of the organism. Variants prove successful if the extrasomatic environment favors them in competition over unaltered meristems. If a variant cell line does not prove capable of, say, attaining a sufficient growth rate, then another meristematic cell may begin growth and overtake the variant. If the variant is not successful, the loss to the organism is minimal; only a few cells are lost and a new lineage can be quickly established from the totipotent line. In this manner, as Whitham and Slobodchikoff[11] have recently emphasized, plants may "track" environmental variations with potentially great precision with neither the considerable costs associated with transgenerational modes of evolutionary change (e.g., the cost of dispersal), nor the threat of somatic cell parasitism.

The rigidity of cell walls, however, limits the free movement of nutrients from one location to another. For an organism with rigid cell walls, there are only two ways in which the benefits of cell movement may be retained. Either specialized conducting tissues must arise, as occurred in vascular plants, or cell division must be decoupled from nuclear division, producing a multinucleate, coenocytic entity. It is this coenocytic condition that characterizes the Kingdom Fungi. The coenocytic condition allows free movement of nuclei via cytoplasmic streaming from one part of the mycelium to another. In particular, nuclei and organelles can be moved across the common cytoplasm and deployed at those locations most beneficial to the organism as a whole. Inherent, however, in the ability to move the cytoplasm about freely is the potential liability that organellar and nuclear variants have access to both energetic resources and sites of production of germinative tissues.

Mycologists have been far more attentive to the problems of suborganismal variation and its consequences than have been either botanists or zoologists. The propensity, and in some species the necessity, for fusion of genetically distinct

11. Whitham, T. G. and C. N. Slobodchikoff. 1981. *Oecologia* 49:287–292.

mycelia has resulted in a wealth of empirical studies on the influences of genetic variation within mycelia. Pontecorvo, in 1946, advocated an approach to variation within mycelia that this book attempts to resurrect and extend to all multicellular organisms:

> We may be justified in considering a hypha as a mass of cytoplasm with a population of nuclei. Such a population is subject to: (1) variation in numbers; (2) drift—i.e., random variation in the proportion of the different kinds of nucleus; (3) migration—i.e., influx and outflow of nuclei following hyphal anastomosis; (4) mutation; and (5) selection. There are here all the elements considered by Fisher, Haldane, and Sewall Wright in their work on the genetical theory of populations. No doubt their techniques will be to a large extent adaptable, and extremely useful in, the study of heterokaryotic [chimeric] systems when the time has come for a parallel treatment.[12]

Mycological work since Pontecorvo advocated this approach has clearly demonstrated both that chimeric mycelia are often effective in exploiting environments that one or both of the components of the chimera are unable to exploit in isolation, and that this effect is due to the movement of different nuclei to the site at which their synthetic activity is needed. This same work, however, has equally clearly demonstrated that the benefits of chimerism are not won without the threat of significant losses.

The same free movement of nuclei that allows chimeras to display vigor is coupled with an extraordinary susceptibility of fungi to nuclei and organelles which fail to support necessary maintenance functions. This susceptibility is of paramount importance, as deleterious mutations arise far more frequently than mutations which arise *de novo* and yet improve on an already sophisticated design. Atwood and Mukai, for example, report that of 26 mutants recovered from a single *Neurospora* mycelium, 24 were lethals.[13] In addition to nuclear variants, organellar mutations are also well

12. Pontecorvo, G. 1946. *Cold Spring Harbor Symp. Quant. Biol.* 11:193–201.
13. Similar examples may be found in the work of J. J. Miller (1945, 1946a,b. *Can. J. Res.* C23:16–43, 24:188–212, 213–223) and G. W. Fischer (1940. *Mycologia* 32:275–289).

known. Several of the principal mutations exploited in investigations of mitochrondria biosynthesis, for instance, were first recovered from fungi. Perhaps the most dramatic example of such detrimental variants is the observation that naturally occurring mycelia of the devastating rice pathogen, *Fusarium fujikuroi*, harbor a substantial proportion of nuclei which lack the capacity to invade rice.[14]

The coenocytic condition allows the benefits of movement precluded by cellularization, but at the cost of harboring detrimental variants. Increasingly sophisticated controls over this variation are a principal theme in fungal evolution. In many of the most primitive fungi the cell cyles of all nuclei in the mycelia are synchronized. Nuclei divide in unison. While cell-cycle synchronization does not control the rate at which detrimental variants arise, nor their access to sites of reproductive activities, this trait nevertheless has a profound effect on the potential for parasitism.[15] If all nuclei divide as one, no variant line can increase at a rate greater than the nuclei from which it arose. An important class of potential variants is thereby rendered ineffective. A variant which fails to, say, synthesize necessary products gains no advantage over nuclei that perform this function in terms of their relative rates of division. With cell-cycle synchronization, variants may arise and may be inherited, but are unable to monopolize a mycelium.[16]

14. Ming, Y. N., P. C. Lin, and T. F. Yu. 1966. *Scientia Sinica* 15:371–378.
15. Cell-cycle synchronization is not limited to fungi. The problem arises in any organism in which nuclear division becomes decoupled from cell division. For example, the myxomycetes produce a streaming plasmodial stage composed of many thousands of nuclei, all of which divide in synchrony. The embryos of many cryptograms and some higher plants (e.g., *Ginkgo* and most gymnosperms) have a brief coenocytic stage in which all nuclei are synchronized. The same occurs in animal embryos which are temporarily coenocytic (e.g., *Drosophila*). The evolution of cell-cycle synchronization, to equalize rates of replication among nuclei of a common cytoplasm and thus limit the proliferation of variants, is probably an ancient trait.
16. A device analagous to cell-cycle synchronization occurs in some colonial tunicates (e.g., *Botryllus*), which are composed of a series of asexually budded zooids joined by a common vascular system. Here sexual reproduction and asexual budding alternate through a regular series of complex morphogenetic movements. These movements are synchronized throughout the colony, ensuring that no variant lineage is likely to proliferate disproportionately. Developmental staging events in metazoans may well represent the cellular analog of cell-cycle synchronization in coenocytic organisms.

Cell-cycle synchronization is not ubiquitous in the fungi. In many forms, septation is found. Septal units are joined by pores. Pore structure displays considerable variation; some actively limit nuclear migration between continuous septal units, while others allow considerable movement (Figures 4.1, 4.2). The control of nuclear movement may display amazing subtlety. Raper and San Antonio, for example, have found that when chimeric mycelia are produced which harbor two distinct nuclear populations, one of which has a pronounced deficiency in synthesizing certain important metabolic products, but with no deficiency in reproductive ability, septa act to selectively allow the migration of normal nuclei across septal borders while denying migration of defective nuclei.[17] Indeed, the extent and timing of septation in chimeric mycelium is under precise genetic control.[18]

Cell-cycle synchronization and septation, however, are merely preludes to a far more effective control over variation, that of controlled cellularization. All fungi cellularize at one point in ontogeny—when reproduction occurs. The cellularization may be mycelium-wide, but is more commonly limited to the site of reproductive activity. Although this cellularization must in part reflect the necessity of compartmentalizing a small quantity of cytoplasmic, nuclear, and organellar material for dispersal, the extent of cellularization in all forms is far greater than that required for this task alone. Cellularization prior to reproduction acts to limit the free access of nuclei to the site of reproduction. Any variant which fails to attend to supportive functions, but rather drifts in the cytoplasm seeking sites of reproduction, will be discouraged.

In the most advanced and highly septate of the fungi, the basidiomycetes, a particularly specialized form of cellularization has arisen whose function is wholly to mediate the distribution of genetically distinct nuclei within the mycelia. Many basidiomycete species require the fusion of genetically distinct mycelia prior to sexual reproduction. When such fusion occurs, a structure is differentiated that was previously lacking. At the point of hyphal fusion, clamp connections are formed (Figure 4.3). Clamp connections link a chimeric cell with a homogeneous one and act to transfer a single nu-

17. Raper, J. R. and J. P. San Antonio. 1954. *Amer. J. Bot.* 41:69–86.
18. Raper, J. R., *The Genetics of Sexuality in Higher Fungi*. New York: Ronald Press, 1966.

FIGURE 4.1

The structure of septa
pores in various fungi.
(From Burnett, 1976.)

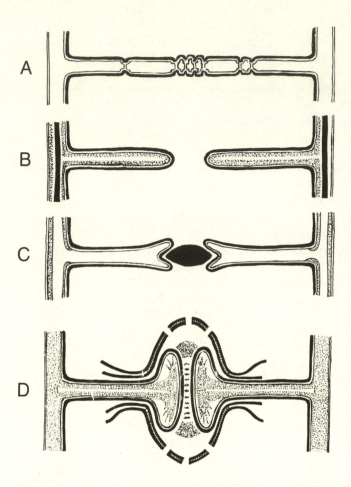

cleus to the latter cell, rendering it chimeric. In this fashion,
cell-by-cell, the two mycelia eventually become chimeric
with no single cell represented by more than one nucleus
from either component of the chimera, ensuring that nuclei
do not have free access to sites of reproduction.[19]

The ancestors of the plant, animal, and fungal kingdoms
faced the same conflict between selection at the level of the
individual and at the level of the cell lineage, but differed
with respect to their ancestral states of cellularization and ar-

19. The "clamp connections" of basidomycetes are homologous with the
pattern of cellularization which occurs just prior to reproduction in ascomycetes. Both serve the same function, that of assuring that nuclei do not
have free access to the sites of reproductive activity.

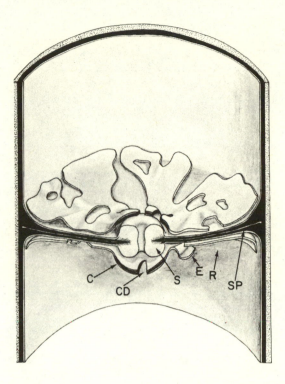

FIGURE 4.2

A three-dimensional reconstruction of the septal pore structure common in the basidio-mycetes (portrayed in cross-section in Figure 4.1D). (From Bracker and Butler, 1963.)

chitecture of cell membranes. These initial states defined the susceptibility of each clade to the effects of variants which failed to support organismal function, and ordained that development would follow different evolutionary pathways in each group. Animals, with non-rigid cell membranes, were enormously susceptible to such variants and evolved exceedingly complex programs of early embryonic determination and morphogentic movements to bring differentiated cells into play at the appropriate place and time in ontogeny. Plants, by virtue of their rigid cell walls, were far less susceptible to effects of detrimental variants and, accordingly, lack the developmental complexity that metazoans evolved to control such variation. Fungi, coenocytic with rigid walls, were burdened with the limitations of both plants and animals: their coenocytic condition allows free movement of variants through a common cytoplasm, and their rigid hyphal walls preclude substantial movement. The most advanced fungi have come to realize the conditions which were

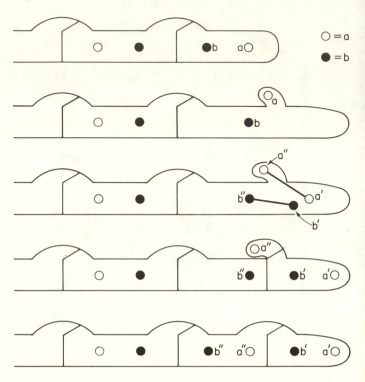

FIGURE 4.3   Stages in the formation of clamp connections, illustrating the process by which different nuclei (a, b) become evenly distributed in each septal unit. (From Moore-Landecker, 1972.)

primitive to the plants—they have become rigid-walled and cellularized.

IV

Who has not felt alternately bewildered and fascinated by the complex schematic representation of the life cycle of some new organism, following round-and-round the various byways in attempt to gain an intellectual foothold? The simplest life cycles are little more than alternating progressions of sex and development, the features with which we have previously dealt. The ubiquity of sex and the various developmental modalities which evolved so differently in response to differing ancestral specializations are, taken alone, a poor guide to the complexity of many life cycles. They represent only one thin line—albeit the crucial line linking one meiotic event to the next—among many. Intercalated on that same diagram may be numerous other pathways that beg for explanation.

The developmental modalities of modern kingdoms are held here to reflect innovations in the face of the primitive constraints of cellular architecture. The evolution of development is the generation of a "somatic ecology" that mediates potential conflicts between the cell and individual, while the organism is simultaneously interacting effectively with the extrasomatic environment. The two ecologies, somatic and extrasomatic, are not fully decoupled. The extrasomatic ecology may make non-negotiable demands, to which the ancestral somatic ecologies of each clade may differ in their capacity to respond. To the extent that similar responses are possible in differing developmental contexts, broad phylogenetic similarities in life cycles, independent of developmental mode, may be expected to emerge. Likewise, the somatic ecology may prove inflexible, and the extrasomatic environments available to that clade may become restricted. It is against this backdrop that restrictions on phylogenetic distribution of life cycles arise. Finally, the particular innovations each somatic ecology has made in response to extrasomatic demands may, in turn, provide opportunity for further modifications in development to exploit new extrasomatic environments. Such opportunities will ultimately be reflected in unique, idiosyncratic, life cycles correlated with particular developmental innovations. Herein lies the potential to begin to formulate a workable theoretical structure in which to interpret the diversity of modern life cycles: any given life cycle trait will be differentially accepted, rejected, or modified as a function of the somatic context in which it arises under the inflexible rule that the developmental program must nevertheless maintain controls over potential conflicts between selection at the level of the cell and of the individual.

*

While the path from zygote to adult is the sole mechanism for establishing a new *genetic* individual, many life cycles include a pathway for the production of new *physiological* individuals of identical genetic composition by asexual means. A microscopic segment of a fungal hypha, a bit of the protenema of a moss, or a dislodged fragment of sponge may bypass the entire embryology orchestrated by a zygote and, nonetheless, come to establish a physiologically discrete new individual.[20] No less striking are asexual dispersal products (Figure 4.4), seen in condial spore of a fungus, the statoblast of a bryozoan, or the gemmae of a bryophyte. Both avenues represent a primitive capacity to package a totipotent cell, with or without associated somatic lineages, competent to create a new individual without passing through the myriad complexities of each ontogeny.

The ecological and genetic benefits of this life cycle modality are many.[21] A successful clone may spread without diluting a particularly favorable genetic association by meiosis. A clone may reproduce without the necessity of locating a mate, nor the cost of contributing half its reproductive efforts to the production of males. A clonal mode of propagation allows rapid proliferation in response to newly located resources or equally rapid expansion in a brief seasonal window free of predators. Packaging of different-sized asexual propagules, each with a distinct dispersal system, allows the exploitation by a given clone of resources distributed in peculiar but nonetheless stereotyped ways. This diverse array of potential benefits has formed the basis for considerable discussion of the manner in which particular ecologies may favor the adoption of clonality.

20. The fact that such fragments are capable of orchestrating the production of a new individual without passing through the various stages displayed by the developing zygote has traditionally been regarded as something of an enigma. If the organism is capable of organizing itself into a discrete physiological unit from a mere mass of dislodged cells, why is the embryonic progression so stereotyped? The clue to this phenomenon may well lie in the arguments presented previously. The metazoan zygote follows a path enforced upon it by its mother, or by other developing cell lineage(s), because differentiation of cell lineages in early ontogeny often requires the lineage to limit its own replication. An embryonic lineage will stop replicating only if the mother or another cell lineage forces it to do so. Such considerations are removed in the case of the asexual fragment. Cell lineages have already become determined in earliest ontogeny, and a full complement of somatic stem cells is provided in the propagule. The propagule lacks an embryology because the conflicting forces shaping embryology—conflicts between selection at the level of cell lineage and at the level of the individual—have long since been resolved.

21. Detailed discussion of the genetic and ecological correlates of clonality may be found in Bell's monograph (*op. cit.*) and in a recent symposium (J.B.C. Jackson, L. W. Buss, and R. E. Cook, eds. *Population Biology and Evolution of Clonal Organisms*. New Haven, Conn.: Yale University Press, 1985).

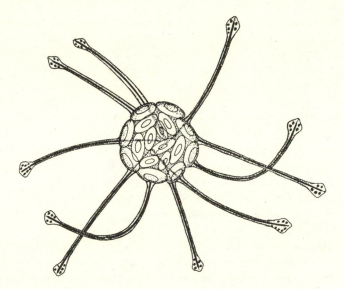

FIGURE 4.4   An external asexual bud of the sponge *Alectona*. (From Garrone, 1974.)

The phylogenetic distribution of clonality (patterns in the long list of organisms which display it and organisms which lack it) does indeed map in an appealing fashion on ecological circumstance. Yet it is not clear here what is cause and what is effect. Primitively clonal organisms surely display asexuality, in part, because their ecologies have favored its retention. In Van Valen's oft-quoted terms, "evolution is the control of development by ecology."[22] However, the absence of asexuality in a life cycle may have a quite different explanation than does its presence: if a developmental innovation in any way precluded asexuality, the absence of the trait reflects less an absence of an ecology which favors the trait than the molding by development of an ecology that can persist without it. Evolution may also reflect the control of ecology by development.

In the histologically least elaborate of the metazoan taxa, the Porifera and the Cnidaria, somatic cells adopt a single state of differentiation; if they divide at all, they divide to give rise to a cell in the same state of differentiation (Figure 3.14). All somatic cells initially arise from a totipotent cell, competent to give rise not only to all somatic cells but also

22. Van Valen, L. 1973. *Science* 180:488.

to give rise to the gametes. It is these same cells, if properly packaged, that give rise asexually to a new individual. A major developmental innovation in the metazoa, so crucial for the subsequent evolution of germ-line sequestration and stabilization of the epigenetic programs of higher metazoans, was the replacement of this totipotent cell type by secondary somatic differentiation. With the capacity for some somatic cells to give rise to other somatic cells in a quite different state of differentiation, the potential arose for totipotent germ-line cells to abandon mitotic activity and to become sequestered from the task of continuously replenishing depleted somatic stocks, effectively reducing the potential for variants to arise in the course of ontogeny and prove heritable.

Germ-line sequestration was purchased at a cost. Without a mitotically active totipotent cell lineage, asexual reproduction becomes impossible. Asexual reproduction requires a continued supply of totipotent cells to be packaged in propagules, while germ-line sequestration requires that cells totipotent to produce a new individual be set aside in early ontogeny. No organism could possibly sequester in earliest ontogeny all those cells required to propagate asexually throughout a potentially long life span.[23] The only alterna-

23. Some organisms which reproduce clonally live a remarkably long time. Exclusively sexual organisms, by comparison, are short-lived. Aspen clones in Colorado are said to have flaunted their brilliant annual displays no fewer than 10,000 times. Similar figures are known from creosote bushes in the Mojave Desert (>11,000 yrs) and huckleberry clones (>13,000 yrs). Many coral colonies may well have lived through the last ice age. The life cycles of aspens, creosote, huckleberry, bracken, and corals share one important characteristic—all are capable of asexual propagation. Is this a fortuitous association, or are asexual organisms potentially immune to the physiological degradation waiting for all of us? The brevity of the life span of asexual insects or many annual plants is ample evidence that asexuality *per se* does not ensure longevity; yet, the fact that the longest-lived organisms are disproportionately organisms capable of asexual propagation must certainly give pause.

Longevity and senescence were first treated as evolutionary issues by August Weismann, who wrote: "Death takes place because a worn-out tissue cannot forever renew itself. . . . Worn-out individuals are not only valueless to the species, but they are even harmful, for they take the place of those which are sound" (*Essays on Heredity and Kindred Biological Problems*. 2nd ed. E. B. Poulton, S. Schonland, and A. E. Shipley, eds. and trans. Oxford: Clarendon Press, 1891:24). Weismann's contribution is valuable more for his perception of natural death as a phenomenon shaped by natural selection than for his interpretation of it. As Medawar points

tive would be for the totipotent lineage to remain mitotically active, to reverse the very process of sequestration.
Metazoans with a sequestered germ line do not reproduce
asexually.

The non-rigid cell membranes of metazoans provided op-

---

out, Weismann's explanation for the evolution of senescence is circular:
"By assuming that the elders of his race are decrepit and worn out, he
[Weismann] assumes all but a fraction of what he has set himself to
prove." (*The Uniqueness of the Individual*. London: Methuen, 1957:3).

Medawar replaced Weismann's flawed arguments with a widely accepted explanation of the evolution of senescence (*ibid.*). Medawar based
his argument on an analogy to a population of potentially immortal, sexually reproducing, test-tubes. These test-tubes could be destroyed, say by
dropping them on a laboratory floor, but were potentially immortal in the
sense that they did not degrade with time. If carefully handled, they
would persist forever. He then asked about the fate of a population of such
test-tubes. In routine laboratory operation, some test-tubes would inevitably be accidentally destroyed. As time passed, the number of older test-
tubes would necessarily decline—not because they had aged and were
therefore decrepit, but simply because they had been exposed to the constant danger of human clumsiness for a longer period.

The attenuation of the test-tube age-frequency distribution, resulting
from a constant risk of mortality over long intervals, is crucial. Since older
test-tubes are few in number, they contribute only a small fraction of offspring to the next generation. Hence the force of natural selection declines
with increasing age. Medawar animates his simple model by asking the
reader to consider the fate of a test-tube gene which causes individuals carrying it spontaneously to burst asunder at great age. Such a gene will remain in the population, for the individuals which suffer its effects are few
and those few who do suffer have already produced the vast bulk of the
offspring that they would have produced during their entire life span had
the gene-product not struck them down. The cumulative effect of many
such genes, especially a subset of them which display a pleiotropic beneficial effect when the organism is young and a detrimental effect when it
is old, will inevitably be a period of gradual decline and dysfunction with
increasing age.

Medawar's test-tubes reproduced sexually; each new test-tube began
life at age zero. Medawar's argument may be fruitfully extended to populations in which new individuals arise both sexually and by asexual
means. Asexual individuals differ from sexual individuals in that an
asexual propagule begins life at *physiological age* zero, but at a *genetic age*,
calculated from the last occurrence of sexual reproduction (that is, zygotic
age). An asexual offspring has a genetic age equivalent to that of its parent, and its asexual offspring produce asexual offspring of the genetic age
of their grandparents, and so on. Quite modest rates of asexual reproduction can produce quite dramatic effects on the age-distribution of the resulting population. While the age-frequency distribution of an asexually
reproducing organism eventually attenuates just as that of a sexual organism will, this attenuation will occur at a considerably greater age than it

portunities for variants to arise and move into the germ line. Adaptation to internal control of variants established germ-line sequestration, and germ-line sequestration removed the option of asexual reproduction. The limitation that germ-line sequestration places on life cycle evolution, however, applies only to animals. Asexual reproduction is common in the life cycles of many fungi and plants. Both groups are characterized by rigid cell walls, and are constrained by this architecture to retain an active totipotent line. The primitive condition of plants and fungi precluded cell movement, germ-line sequestration did not evolve, and asexuality remains a viable means of reproduction in at least some members of every major plant and fungal higher taxon.

The limitation placed on asexuality in metazoans is not absolute; a secondary loss of sexuality can reestablish asexuality in the life cycle. Only a relatively small number of taxa have done so, despite the well-known "two-fold cost of sex" (Table 4.2). The limited extent of secondarily derived asexuality is not surprising. An organism with germ-line sequestration can reproduce asexually only by modification of one stage in ontogeny—in that brief window between fertilization and the terminal determination of the germ line. Indeed, asexuality, when it occurs, is derived from multiple cleavage of a fertilized egg, activation of the egg nucleus without fertilization, or via unusual patterns of inactivation of one chromosomal complement following fertilization (Table 4.2). The secondary acquisition of asexuality does not restore the capacity for unlimited replication of totipotent cells; it occurs via evolutionary modifications expressed in the brief ontogenetic interval between activation of the egg nucleus and germ-line sequestration. This limitation is revealing. Those metazoans which have secondarily acquired asexuality can only produce the same number of asexual off-

---

would in an otherwise similar sexual population. Asexually reproducing individuals are not immune from the effects of aging, but are potentially far less susceptible to early aging than sexual organisms. While aspen, fairy-ring fungi, and reef corals pass millennia, humans, as organisms whose remote ancestors adopted germ-line sequestration and abandoned asexuality, are destined to die at relatively early age. Germ-line sequestration not only closed off asexuality as a developmental alternative, it limited the organisms possessing it to only a brief span of life. (I first presented this argument at a recent symposium on clonality, at which Hal Caswell presented an independently derived, analogous argument. [Jackson, Buss, and Cook, *op. cit.*, pp. 187–224].)

TABLE 4.2   Metazoan Secondary Asexuality[1]

Mode	Taxa	Cytological Characteristics
Polyembryony	Scyphozoa, monogenean trematodes, loxosomatid entoprocts	Embryo divides during early development to give rise to several identical offspring
Apomixis	Digenean trematodes, bdelloid rotifers, some nematodes, tardigrades, coccids, cladocerans, aphids, cynipid wasps, teiid lizards	Unreduced eggs develop without fertilization
Arrhenotoky[2]	Monogonont rotifers, most Hymenoptera, some mites, beetles, and homopterans	Meiosis without syngamy in males; females produce reduced eggs which develop into diploid females if fertilized and haploid males if not

1. Modified from G. Bell, *The Masterpiece of Nature*. University of California Press, 1982.
2. Partial mixis, thus not true asexuality.

spring (or, in the case of polyembryony, a modest increase in number) as they can sexual offspring. Ecology may reassert control over development, but only in a brief window of ontogeny in organisms whose environmental context permits a modest asexual reproductive output. It is hardly surprising that secondarily asexual taxa are virtually all small creatures (Table 4.2).[24] Small organisms can complete their life cycles quickly, and thus are capable of rapidly exploiting a new or seasonally constrained cache of resources despite the inherent restriction that development has placed on their ecology.

The adoption of a secondarily asexual state, while severely constrained, nevertheless holds potential for a class of further life cycle evolution. The epigenetic programs of ancestral metazoans arose as variants in, or gaining access to, the to-

24. Rare instances of relatively large secondarily asexual organisms are known (e.g., *Poeciliopsis*, a Mexican fish). Such organisms inhabit refugia, free from predators which feed upon their young in more typical habitats. This habitat restriction is revealing, for they live in precisely those habitats in which the impact of the developmental restriction on reproductive output is minimized.

tipotent heritable cell lineage. With the adoption of germ-line sequestration, further heritable variation was rendered very unlikely. However, secondarily derived asexuality, when it alternates with a sexual phase, provides a mechanism by which variation arising in the course of ontogeny may again become heritable in the asexual generation and yet spread throughout the population in sexual generations. The small number of cell divisions per generation in the germ line of a sequestered organism is potentially greatly amplified by intercalation between each sexual generation of a number of asexual generations, each of which entails a number of heritable divisions.[25] Metazoans displaying a secondary acquisition of asexuality often display extraordinarily complex life cycles in which the organism undergoes a series of morphologically distinct phases, the transitions between which are punctuated by episodes of asexual reproduction.

This phenomenon may be expected to hold particular significance for organisms in which each asexual phase is challenged by a unique environment, as is common in the life cycle of metazoan endoparasites. Digenic trematodes, for example, exhibit a complex life cycle involving several asexual generations interspersed between each sexual generation. Each stage is represented by its own distinct larval form (Figure 4.5). The hermaphroditic sexual fluke produces a zygote which divides into one large somatic cell and a smaller germinative cell. The germinative cell continues to cleave for a period, generating new somatic lineages, after which it retires to the posterior of the embryo while the somatic lineages elaborate the miracidium larvae. The free-swimming miracidium locates a host, generally a mollusc, and bores into its tissues. Upon entering the host, the somatic tissues of the miracidium are reduced, eventually producing a sporosac consisting of outer epithelial and muscular layers and a hollow interior housing the germinative cells. The sporosac stage moves throughout the host, absorbing nutrition through its epithelium, while the germinative cells give rise to germ balls, which are, in reality, asexually produced embryos. From these germ balls and germinative cells arises the redia, which escapes from the sporosac to further penetrate the host. The redia arises in the same manner

25. Perhaps no less important than the increase in absolute number of heritable cell divisions is the fact that asexuality reinitiates the single-celled state.

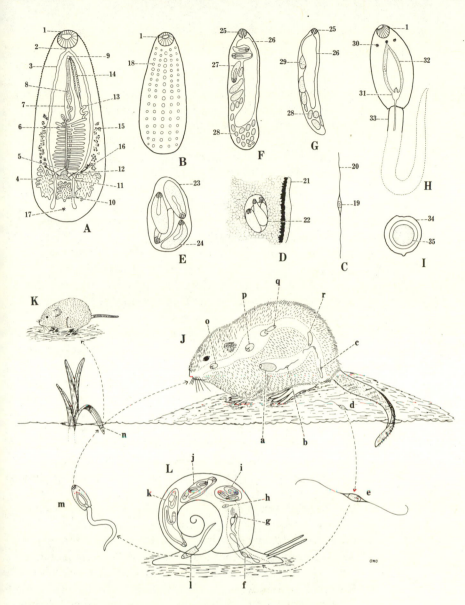

FIGURE 4.5   Life cycle of the digenic trematode *Quinqueserialis quinqueserialis*. Adult worms (A, B, a), parasitic on the muskrat *Ondatra zibethica* (J) or on the meadow vole *Microtus pennsylvanicus* (K), shed eggs (C, b–f) which are eaten by the snail *Gyraulus parvus* (L). The eggs hatch in the snail's mantle (g) and the miracidia penetrate the gut (h) to give rise to sporocysts (D, E, i). The sporocyst gives rise to the redia (F, i), which migrates to the liver. Mother rediae give rise to daughter rediae (j), which in turn give rise to cercariae (G, k). Cercariae (H) escape from the snail (l–m) to encyst upon vegetation (I). Muskrats eating the vegetation become infected (o–p). (From Olsen, 1974.)

as does the miricidium from the zygote: unequal cleavage of germinative cells leads to somatic lineages which elaborate the redial organs and to further germinative cells which remain quiescent. The germinative cells, or the germ balls, of the redia subsequently proliferate to produce yet further asexually derived redial generations and ultimately a quite different, but also asexual form, the cercaria. The development of the cercaria is again similar to that of preceding stages, with germinative cells, this time, set aside for the production of the sexual generation. The cercaria escapes the host tissues to adopt a brief free-living existence. The cercaria may be consumed directly by its definitive host or may encyst as a juvenile fluke (the metacercaria stage) until such time as it is consumed. The adult fluke feeds on this new host, eventually producing gametes and reinitiating the cycle. The number of cercariae derived from a single zygote varies with details of the life cycle, of which there are many, but may reach one million through sequential asexual amplification in the sporosac and multiple redial stages.

The complex life cycle of metazoan endoparasites such as digenic trematodes reveals the potential latent in the secondary loss of sexuality. Multiple asexual generations are intercalated between each sexual generation. The restriction on the number of heritable divisions per sexual generation introduced by germ-line sequestration is relaxed, once again opening the germ line to heritable variation, and fueling specializations to the particular ecological circumstances faced by the asexual generation. The complex interaction of history with the developmental constraint is evident. The life cycle of digenic trematodes testifies to the importance of actual historical sequence in evolution. A developmental innovation, germ-line sequestration, constrained metazoan life cycles by removing the asexual path. Ecological circumstance favored a return to asexuality in some forms, setting the stage for subsequent elaboration. In forms with multiple asexual generations intercalated between sexual ones, opportunities for variation arising in the course of ontogeny to prove heritable were provided once again, allowing the complicated life cycles of facultatively asexual metazoans to evolve further.

The developmental modalities evolved in plants, fungi, and simple metazoans to regulate conflicts between selection at the level of the cell and at the level of the individual, despite their inherent differences, all permit asexuality. The occurrence of asexuality is not correlated with developmental mode. However, in higher metazoa, germ-line sequestration, a developmental modality crucial to the maintenance of individuality, is at odds with an asexual avenue in the life cycle. Accordingly, the absence of asexuality is highly correlated with the presence of germ-line sequestration. Finally, a subsequent developmental specialization, the secondary adoption of asexuality via modification of the germ line, has made possible many highly idiosyncratic life styles.

*

The origin of individuality by no means ensured its integrity. Those organisms which first evolved cellular differentiation must have enjoyed enormous success. Sedentary individuals may have been so common that they occasionally came to live side-by-side. Herein lies another difficulty. Genetic homogeneity of the individual can be disrupted not only by mutation, but also by the fusion of genetically distinct individuals. Fusion yields a chimera and reinitiates competition between cell lineages. Within populations, individuals vary, often within wide margins, in their respective programs of somatic versus gametic investment. Fusion between individuals with differing investment routines will necessarily result in parasitism of one component of the chimera on the other. Mechanisms to prevent indiscriminate fusion must surely have followed closely on the heels of the evolution of cellular differentiation.

Fusion among colonial metazoans is demonstrably associated with parasitism of one component of the chimera on the other. Müller, working with the colonial hydroid *Hydractinia echinata*,[26] found that subsequent to fusion of male and female colonies, the male component of the chimera always came to dominate the eventual production of gametes from

26. Rejection of incompatible tissue in this hydroid is associated with the induction of a specialized organ of defense, the hyperplastic stolon. The tips of such modified stolons are armed with nematocytes, stinging cells, which act to destroy the tissue of incompatible neighbors. The association of inducible competitive defenses and historecognition is widespread among the Cnidaria (see Buss, L. W., C. S. McFadden, and D. R. Keene. 1984. *Biol. Bull.* 167:131–158). Additional inducible, nematocyst-based, effector systems include the acrorhagi and catch tentacles of sea anenomes and the sweeper tentacles of scleractinian corals. Intercolony competition is similarly coupled with historecognition in the Bryozoa,

the chimeric soma (Fig. 4.6).[27] Similarly, Sabbadin has recorded the disproportionate reproductive success of one component of a chimera in the ascidian *Botryllus schlosseri*.[28] All major colonial metazoan taxa (the Porifera, Cnidaria, Ascidia, and Bryozoa) display genetically based systems of fusion and rejection, whereby fusion is typically restricted

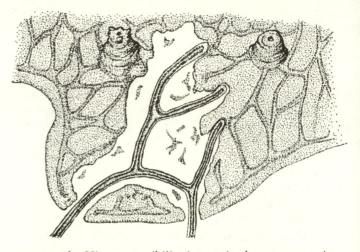

FIGURE 4.6   Histocompatibility interaction between two colonies of the athecate colonial hydroid *Hydractinia echinata*. A stolonal extension from one colony has destroyed the tissue of the opposing colony at all points of contact. Closely related individuals of this same species fuse with one another upon contact. (From Müller, 1964.)

Ascidia, and Porifera, although less obviously associated with inducible effector systems.

The coupling of historecognition with intraspecific competition strongly implies that the fusion/rejection loci of clonal invertebrates are genes which act to control the units of selection. Fusion results in competition between cell lineages, and rejection results in competition between individuals. The decision to fuse or to reject is a decision to compete at the level of the cell or at the level of the individual.

27. The association of sex with the outcome of cell lineage competition in a chimera is revealing (Müller, W. 1964. *Wilhelm Roux' Arch. Entwicklungsmech.* 155:181–268), as it suggests the potential for sexual selection to shape the evolution of historecognition. This suggestion will surely be further fueled by the recent observation of a sexual restriction on the deployment of historecognition effector systems in the anemone, *Metridium senile* (Kaplan, S. W. 1983. *Biol. Bull.* 165:416–418).

28. Sabbadin, A. and G. Zaniolo. 1979. *J. exp. Zool.* 207:289–304.

solely to close kin. This is allorecognition and it serves to limit fusion in organisms whose sedentary habits put them at risk of fusion.

In contrast to the sessile taxa, mobile organisms, which are not at risk of fusion, often lack allorecognition. Tissue grafts can be readily transplanted from one individual to another without any sign of rejection.[29] Multiple attempts to detect allorecognition in annelids, molluscs, nematodes, platyhelminths, and arthropods have failed.[30] As mobile organisms, they simply do not encounter an ecological context in which fusion can occur.[31] The risk of somatic cell parasit-

29. The failure to display allorecognition, that is, the capacity to distinguish between individuals of the same species, does not necessarily imply a lack of xenorecognition, the capacity to recognize different species. For example, insects, which are unable to recognize allografts, are quite capable of mounting a humoral response to bacteria.

30. It is a rather remarkable correlation that metazoans which behave as endoparasites on other metazoans are disproportionately successful on those hosts which lack sophisticated mechanisms of allorecognition. Indeed, major adaptive radiations may have arisen as a consequence of the failure of some taxa to display allorecognition. For example, the entire Order Ichneumonidae, some 40,000 species of wasps most of which parasitize lepidopteran caterpillars, may represent the invasion of an adaptive zone left open by the lack of a recognition system. Even those parasites which have succeeded in invading hosts with elaborate immune systems typically do so only after a primary, and presumably primitive phase in an invertebrate host. A detailed comparison of the major adaptive radiations of multicellular endoparasites and the phyletic distribution of allorecognition may prove quite illuminating.

31. The vertebrates (and the echinoderms) represent obvious exceptions to the generalization that allorecognition systems occur in sedentary taxa for which the threat of somatic cell parasitism is demonstrably substantial. As Medawar realized (*op. cit.*), the occurrence of transplant rejection (i.e., self-recognition) in vertebrates is a genuine evolutionary enigma. Why should each of us carry around in our genome a huge gene complex whose function is the prevention of intraspecific fusion, an event which will surely never occur naturally?

Two hypotheses have been offered to account for this fact (e.g., see Buss, L. W. and D. R. Green. 1985. *J. Comp. Dev. Immunol.* 9:191–201). The vertebrate immune system may represent a convergent evolution of allorecognition, one which arose as a nonadaptive byproduct of sophisticated modes of xenorecognition. Alternatively, the immune systems of vertebrates and echinoderms may be homologous with those found in clonal invertebrates, only to have subsequently become adopted as a mechanism of xenorecognition. The latter hypothesis is supported by the fact that several primitive echinoderms (which are presumed ancestral stock for the chordates) were sedentary, potentially clonal, organisms. Thus, vertebrate ancestors may well have encountered fusion as a naturally occurring event and developed allorecognition as a response to the threat

ism which fusion entails does not arise and they lack the adaptation that serves to limit this risk.[32]

Mobility, however, is not an alternative open to plants. Yet members of the plant kingdom display no obvious mechanisms of historecognition. Bormann, for example, found that genetically distinct pine trees, over a considerable area, were fused one to another via their root systems.[33] Indeed, the ease with which plants accept grafts is the basis of a large industry producing seedless citrus fruit. However, the primitive state of the cell wall of plants, just as it limits the effects of plant tumors, similarly protects plants from parasitism following fusion between unrelated individuals. The rigid cell walls of plants restrict cell movement and, accordingly, also ensure that the cells of one component of the chimera cannot move into positions allowing them to gain access to reproductive organs supported by the soma of the other component.[34] Selection at the level of the cell lineage, in the form of somatic cell parasitism following fusion, is not

---

of somatic cell parasitism following fusion. Further support is found in the fact that vertebrate recognition of foreign tissue still requires simultaneous self-recognition (i.e., antigens presented on macrophages result in the release of interleukins only to T-cells which match the antigen in the context of appropriate self markers). Hence the primitive system, though no longer required for fusion once mobility was acquired, was nevertheless required in xenorecognition and, accordingly, not lost in the course of evolution. Both hypotheses are defensible, and the matter will perhaps be settled in the relatively near future, as attempts to locate invertebrate DNA sequences homologous to vertebrate immune-system genes come to fruition.

32. Some primarily mobile taxa have some sedentary representatives (e.g., bivalves, barnacles, and serpulid worms) and accordingly may be suspected to be faced with opportunities for fusion. Yet, the sedentary members of each such group are surrounded by exoskeletons, calcareous tubes, or shells which no doubt effectively mitigate against fusion.

33. Bormann, F. H. In T. T. Kozlowski, ed. *Tree Growth*. New York: Ronald Press, 1962:237–246.

34. Although the rigid cell walls of plants act to prohibit the free movement of totipotent cells from one individual into the soma of another following fusion, a more limited form of "parasitism" may occur. If the graft were to, say, cause nutrients to flow disproportionately to its own tissues or if the flowers of the graft proved more attractive to pollinators and hence limited the access of pollinators to the remainder of the plant, a graft might act to "parasitize" its partner. The rigidity of cell walls does not prohibit competition between the parts of a plant; rather it limits a particular form of parasitism—the direct competition of totipotent cells for access to the germ line.

in conflict with selection at the level of the individual, and thus sophisticated devices to limit fusion have not evolved.

The fungi, like plants, have rigid cell walls. Yet fungi, nevertheless, display quite sophisticated mechanisms regulating the fusion and rejection of genetically distinct mycelia. The existence of rejection systems in fungi is no paradox. Fungi are primitively coenocytic; their nuclei float freely within the cytoplasm. Thus, fusion of adjacent mycelia allows the opportunity for the nuclei of each strain to communicate in a now-common cytoplasm. Indeed, fusion may result in a remarkably rapid transfer of nuclei. Rates up to 10.5 mm/hr have been measured following fusion in *Gelasinospora tetrasperma*.[35] Nuclear migration establishes a novel selective milieu within the now-chimeric mycelium, with the attendant potential for parasitism. Parasitism indeed occurs, as Pittenger and Brawner have described in *Neurospora*, where if one nuclear type is allowed to donate more than 30% of the total nuclear population of the chimera, the donor will quickly come completely to eliminate the host nuclei from reproductive activities.[36] Thus, despite their rigid walls, fungi are as susceptible as sessile metazoans to fusion and the threat of parasitism it entails. Accordingly, fusion-rejection systems are ubiquitous in fungi.

Mechanisms of recognition among the fungi have served to fuel the remarkable diversity of life cycles in this group. Mechanisms which allow fusion are based on an assessment of the degree of genetic similarity of the hypha encountering one another. In many fungi, particularly the ascomycetes, fusion is allowed if both mycelia match alleles at fusibility loci. Fusibility loci are typically characterized by numerous alleles, such that fusion by matching alleles effectively limits fusibility to close kin. Fusibility limited to close relatives, however, is not ubiquitous among fungi. In most basidiomycetes, the criteria which permit fusibility are the opposite of those seen in other fungi. Fusion occurs if alleles at fusibility loci *do not* match. Here fusion is allowed only between unrelated individuals. The two systems have dramatic consequences for the impact of somatic cell parasitism following fusion and the role of fusion in the life cycle of the species.

35. Dowding, E. S. and A. Bakerspigel. 1954. *Can. J. Microbiol.* 1:68–78.

36. Pittenger, T. H. and T. G. Brawner. 1961. *Genetics* 46:1645–1663.

FIGURE 4.7

Fusion between myce-
lia derived from differ-
ent spores of *Coprinus
sterquilinus* on a horse
dung ball. (From
Buller, 1931.)

Fusion has potential benefits, as well as potential costs.[37]
Fusion results in an immediate increase in size (Figure 4.7).
Small organisms are more likely to be eaten, are less capable
of defending themselves, and are often least equipped to
compete with others. An increase in size surely acts as a
buffer against a wide variety of size-related sources of mortality.[38] The timing of first reproduction, an important de-
mographic character in seasonal environments, is typically
size-dependent. Fusion no doubt hastens its onset. Not only
is there an increase in size accompanying fusion, fusion also
results in an increase in the store of genetic variability upon
which the chimera might draw. Chimeric vigor, whereby a
chimeric mycelium is capable of exploiting environmental
conditions which cannot support the growth of either component of the chimera individually, is well known. In haploid organisms, an additional benefit surely occurs. Fusion
buffers the organism against any potentially lethal defect in
one component of the chimera, mimicking the protection afforded by diploidy. Against the ledger of these potential
benefits, however, we must tally the potentially devastating
consequences of parasitism.

Recognition systems limiting fusibility to close kin preserve the best of both worlds. Close kin share a large percentage of their genetic endowment; thus, fusion between

37. See Buss, L. W. (1982. *Proc. Nat. Acad. Sci. USA* 79:5337–5341) for
a review of the potential costs and benefits of fusion.
38. The possession of a recognition system, primitively adapted to control
potential parasitism, may be modified to suit a variety of distinct ecological circumstances. For example, I have argued elsewhere (*ibid.*) that the
association between the benefits of fusion and size might select for delayed
expression of historecognition, such that many closely related juveniles
may fuse to immediately form an association large enough to escape the
mortality bottleneck at small sizes. This, indeed, appears to be the case,
as the hermatypic coral, *Pocillopora damicornis*, has recently been reported
to do just this (Hidaka, M. 1985. *Coral Reefs* 4:111–116). The similarity
of this metazoan life cycle to that of slime molds, myxobacteria, and one
peculiar ciliate is a striking example of evolutionary convergence.
Conversely, an earlier onset of expression of historecognition loci, in
the larval stage, may prove adaptive. The cheilostomatous bryozoan,
Bugula, shows a strong density-dependence in mortality, with solitary
colonies much more vulnerable than those occurring in aggregations
(Buss, L. W. 1981. *Science* 213:1012–1014). M. J. Keough (1984. *Evolution* 38:142–147) has recently demonstrated that *Bugula* settlement is
enhanced in the presence of sibs relative to that realized in unrelated colonies, strongly suggesting that historecognition has been utilized as a
mechanism for habitat selection.

close kin, even if one relative does completely parasitize the other, will nevertheless result in the propagation of a large number of the genes of the parasitized individual. Just as Haldane should have been content "to give his life for two sibs or eight cousins," a fungus should be equally willing to fuse with its close relatives, for fusion with close kin diminishes the potential costs of fusion while preserving its benefits. Indeed, it is not only fungi that exhibit recognition systems which limit fusion to close kin; all known metazoan historecognition systems operate in this fashion.

As mentioned earlier, there are fungi which allow fusion only between distantly related individuals.[39] Terrestrial fungi, as sedentary organisms, are at the mercy of available water or wind to disperse their gametes. Just as a principal evolutionary trend in the plants, second only in importance to the development of vascular tissue, was to evolve increasingly sophisticated adaptations to cope with the difficulties of dispersing gametes on land, primitive fungi faced a similar challenge. The coenocytic condition offers new routes to sexuality that are closed to higher plants. Connections between hypha from different mycelia, reaching each other through the interstices of the soil, provide a novel mechanism of exchanging gametes (Figure 4.8). A recognition system which identifies unrelated individuals, as insurance of outbreeding, followed by an initiation of sexuality, resolves a fundamental difficulty in mate location.

This mechanism of mate location, however, invites parasitism in the chimera. Any further production of asexual propagules by the chimera will surely bear a disproportionate percentage of one component of the chimera. A variety of devices limits such parasitism. Hyphal fusion may be limited to apical septal units which, upon fusion, are immediately walled off. Alternatively, in other taxa, nuclear migration is allowed, a dikaryon (i.e., a chimera) is formed, and all asexual propagation ceases. In these cases, further growth may or may not occur, but asexual propagules are no longer formed. Yet another device explored earlier, is that of precise

39. Forms which display recognition systems which match like alleles may also display a mating recognition system in hyphae specialized for reproduction. The occurrence of both forms of recognition in a given species is suggestive. One wonders whether yeast mating type genes, for which cDNA probes are readily available, would show homology to historecognition genes in, say, *Neurospora*.

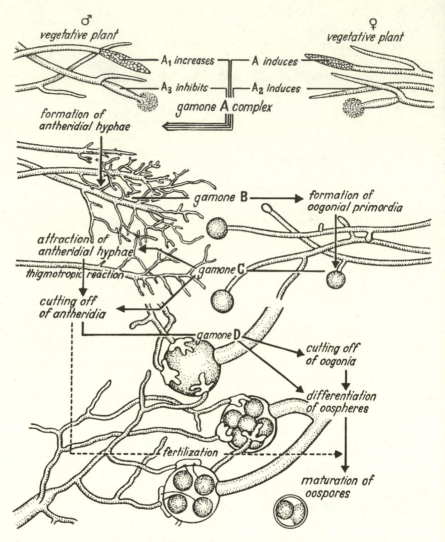

FIGURE 4.8 Stages of the process of mate recognition in *Achlya*. (After Raper, 1955.)

regulation of the nuclear population throughout the mycelium by controlled cellularization, limiting the free access of
nuclei within the chimera. Such forms often continue asexual reproduction, each cell now containing one and only one
nucleus from each partner.

It is not coincidental that the most dramatic fruiting
structures of fungi, the toadstools and mushrooms, are dikaryotic forms in which fusibility is employed in mate recognition. These fungi are completely cellularized. Only via
cellularization can an architecturally sound structure the size
of a toadstool be erected. Not only has this unusual association of sexuality with fusion produced the splendor of mushrooms and toadstools as an artifact of the cellularization it required, it has also provided a mechanism for the evolution of
some extraordinarily complex life cycles. As mentioned earlier, a potential benefit of fusion is that a chimeric mycelium
has the advantage of a greater store of genetic variability than
either component of the chimera possesses alone. The significance of this finding has long been a source of misery to agricultural scientists. In a variety of pathogenic fungi, different strains of the same species often display striking
differences in host preference. Furthermore, heterokaryons
formed between strains with different hosts can often endow
the chimera with a capacity to invade both hosts.[40] Subsequent sexual reproduction between such strains may establish a new race with a wider range of virulence.

The importance of the increased virulence of chimeras is
apparent on inspection of the complex life cycles of rust or
smut fungi. Consider the rust *Puccinia graminis* (Figure 4.9)
as an example. This rust alternately attacks barberry bushes
and wheat plants. The basidiospore, a sexual product of the
heterokaryon, landing upon the leaf of a barberry bush, in-
vades the cells of this host to form a "pycnidium" in the
course of the first growing season. In the following spring,
the pycnidium gives rise to both asexual spores, the pycni-
diospores, and longer, slender, upright hyphae. The pycni-
diospores are exuded in a drop of nectar which presumably
attracts the insects which carry the spores from one pycnidium to another. The spore brought into contact with a

40. For example, E. W. Buxton (1954, 1965. *J. Gen. Microbiol.* 10:71–
84, 15:133–139) has shown that two avirulent strains of the pea-wilt fungus could, as a chimera, becomes as virulent as naturally occurring pathogens.

compatible (i.e., bearing different recognition alleles) hypha fuses with it to form a heterokaryon. The heterokaryon develops into an "aecidium" on the undersurface of the leaf opposite to the position of the "pycnidium." The heterokaryotic mycelium then gives rise to aecidiospores, bearing one nucleus from each of the original strains. This dikaryotic asexual spore is now capable of invading wheat, where it may give rise to yet another dikaryotic asexual spore, the uredospore, which in turn serves as a device to disperse the pathogen to other wheat plants throughout the second growing season. In late summer, the dikaryon produces dikaryotic teleutospores, which are released from lesions on the leaves of wheat finally to undergo nuclear fusion and meiosis. The complex development of this rust on the two host plants is linked at one point by sexuality and at another by the dikaryon (dotted line in Figure 4.9). It is not difficult to imagine that this life cycle arose as a direct consequence of a heterokaryon between two rusts individually capable of attacking wheat and barberry, whose offspring came to sequentially attack both. Indeed, it is difficult to imagine any other scheme by which such a complex life cycle could possibly have arisen.

The forces which spawned the unusual life cycles of fungi are the forces which challenged metazoans to develop elaborate mechanisms of historecognition. Competition within organisms for access to the germ line arises not only through mutation, but also through fusion between conspecifics. The potential consequences of fusion for the integrity of the individual varied as a function of the primitive conditions of the plant, animal, and fungal kingdoms with respect to cell membrane architecture and cellularization. Plants were not challenged. Those animal groups that were challenged responded by evolving sensitive mechanisms of self-recognition. The fungi turned the challenge to their own advantage, inventing from it novel life cycles. A life cycle trait, fusibility, challenged the different clades differently, producing a phylogenetic restriction in those groups which allow fusion. One group, the fungi, responded with a unique developmental innovation which subsequently fueled the evolution of complex, idiosyncratic life cycles.

*

The requirement that evolving life necessarily alternates between vegetative and propagatory phases involves no inher-

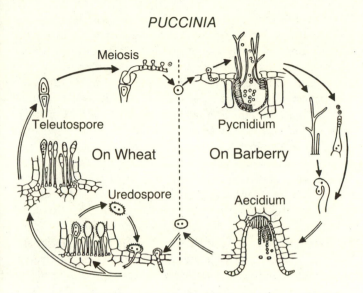

PUCCINIA

Meiosis

Teleutospore

On Wheat

Uredospore

Pycnidium

On Barberry

Aecidium

FIGURE 4.9   The life cycle of the basidiomycete fungus *Puccinia graminus*. Details in text. (From Burnett, 1976.)

ent constraint on the ploidy level of either stage. Yet, ploidy level displays a remarkable phyletic restriction, which surely suggests the operation of strong selection or strong constraint. Animals are overwhelmingly diploid, fungi are overwhelmingly haploid, and plants display a variety of complex alterations of ploidy levels. The pattern is striking, but, curiously, has attracted relatively little attention from evolutionary theorists. The preponderance of diploidy in higher plants and animals has traditionally been attributed to the fact that lethal mutations may be masked in the diploid state. While this benefit is undoubted, it provides little guide as to the phyletic distribution of the trait; no clue is provided as to why fungi and lower plants have retained haploidy.

With the evolution of individuality, selection within the somatic environment became a potentially powerful evolutionary force. The fate of any variant was determined by a new set of criteria. The variant was measured not only in terms of its role in altering the fate of the individual in interaction with the extrasomatic environment, but also in terms of its role in competitive interactions within the somatic environment. The ancestral conditions of cellularization and cell-wall structure, in defining the somatic context in which a variant arises, suggest a mechanism for differential amplification or elimination of variants in each multicellular kingdom. The fate of ploidy changes in the somatic environment merits consideration in this context.

Haploid and diploid cells differ not only in ploidy, but also in size. The transition from the haploid to the diploid state results in a doubling of the haploid cell size. Consider such a ploidy change in the somatic environment of a primitive metozoan. Such a change will be heritable only if it occurs in the totipotent cell lineage. This is the same cell lineage in ancestral metazoans that acts to replenish somatic cells which are incapable of dividing themselves. Herein lies an important fact. All differentiated structures are distance- or volume-constrained; a given distance or volume must be filled with somatic cells if the structure is to be functional. Since haploid cells are half the size of diploid cells, a greater number of haploid cell divisions will be required to produce a given somatic structure than would be needed for production of the same structure from a diploid stock. By virtue of the difference in cell size alone, a diploid cell arising in the course of ontogeny will enjoy a competitive advantage over haploid cell lines within the somatic environment and will increase in relative frequency with time over co-occurring haploid cell lineages. In metazoans, diploidy is favored over haploidy in selection at the level of the cell lineage.[41]

41. Douglas R. Green has suggested a complementary hypothesis, operating at the level of the genome, for the absence of haploidy in animals. He argues that the basis for the competitive superiority of any given cell lineage within the somatic environment is potentially a product of the expression of sequences coding for growth-factor receptors. If the rate of proliferation of such sequences within the genome is less than the rate of deletion of such sequences, then haploid cells would be more likely to lose any basis for superiority in cell lineage competition that they might acquire.

Indeed, a variety of hypotheses are possible. The increased surface area of diploid cells likely results in an increase in the absolute number of growth-factor receptors arrayed on its surface, relative to the number exhibited by haploid cells. Unless the effector mechanism of the growth factor is in some way dependent upon cell volume (which increases disproportionately with increasing cell surface area), the increase in cell surface alone may be sufficient to ensure competitive superiority of diploid cells. The central point here is not the validity of a particular hypothesis regarding the effects of ploidy upon selection within the somatic environment, but simply that such considerations are clearly relevant to the eventual resolution of the problem.

The role of interactions between cells differing in ploidy level within the somatic environment of metazoans invites contrast with the fungi. The contrast is telling. Fungi are coenocytic organisms, and thus ploidy change is not associated with change in cell size. The putative advantage of diploidy in a metazoan somatic environment is obviated in the fungal cytoplasmic environment, and fungi are overwhelmingly haploid. The haploidy of many basidiomycete and ascomycete fungi might, however, be attributable to a different factor. The traditional interpretation of the advantage of diploidy, that of the masking of recessive lethals, holds true for fungi which have developed the dikaryotic state subsequent to fusion. This traditional explanation, however, is alone insufficient. Phycomycetes and several ascomycetes do not form dikaryons and, indeed, many basidiomycetes are monokaryotic for extended intervals. Yet these same forms have retained their ancestral haploid state.

In plants, the fate of a ploidy change would be expected to be quite different. While the ploidy change doubles cell size, as it does in a metazoan, this difference does not result in a corresponding competitive advantage for the diploid cell in a haploid somatic environment. The rigid cell walls of plants restrict the development of the diploid lineage to apical growth from its site of origin. The absence of a somatic advantage associated with ploidy change allows the persistence of a vegetatively haploid condition.[42] Far more important than the retention of a haploid phase, however, is the potential for elaboration of a separate diploid phase. Diploidy has arisen in several plant lineages by the failure of meiosis to occur immediately following syngamy. A primitively haploid plant whose gametes fail to divide immediately following syngamy will act to intercalate a diploid multicellular phase in an originally haploid life cycle.

By resetting the life cycle to a single-celled stage, the po-

42. Diploidization might, in fact, prove to be disadvantageous to the individual in a number of contexts. For example, an erect haploid shoot might be physically unable to support a larger diploid stem. It may not be coincidental that many plants with alternating haploid and diploid life phases involve a substrate-bound diminutive gametophyte from which an erect sporophyte arises. The traditional explanation for this pattern, that of a gradual reduction in the gametophytic phase, must compete with the suggestion that the pattern arises from differential preservation of those sporophytic generations which did not depend upon the gametophyte for support.

tential exists for diploid-phase specific variants to arise and permit exploration of different morphological alternatives. Just as secondary adoption of asexuality in metazoans fueled the evolution of novel life cycles by returning the organism to a single-celled state from which new evolutionary avenues could be explored, the lack of any advantage accruing to ploidy level in the somatic environment of plants allowed the elaboration of alternating, divergent haploid and diploid phases. This elaboration may be relatively minor, as in the red alga *Nemalion* (Figure 4.10). Here the mobile male gametes fertilize immobile female gametes borne on the haploid plant. Meiosis does not follow syngamy; rather the diploid sporophyte begins to divide directly atop the gametophyte, forming a short filament. Meiosis occurs after several cell divisions, producing haploid spores which reinitiate the life cycle. In contrast to *Nemalion*, the diploid phase has become progressively elaborated in several groups. Compare *Nemalion* with the brown alga *Nereocystis* (Figure 4.11). The haploid phase of *Nereocystis* is an inconspicuous plant, one of the innumerable small filamentous algae attached to the sea floor. As in *Nemalion*, the male haploid phase releases motile sperm which fertilize an immobile egg borne at the tip of a female gametophytic filament. Syngamy is again not followed by meiosis, but rather a diploid proliferative phase is intercalated. While *Nemalion* produces only a short diploid filament, the diploid phase of *Nereocystis* produces a huge kelp plant which may grow up 35 meters from the sea floor to the ocean surface before finally undergoing meiosis to resume its humble haploid beginnings. In contrast to animals, the absence of a somatic advantage in plant cells differing in ploidy levels permitted the retention of a haploid stage. In contrast to fungi, the cellularization of plants allowed the intercalation of a diploid phase between haploid generations, generating diverse modes of alternating ploidy level and the complex developmental schemes seen in modern plants.

V

With the evolution of individuality, selection on the vegetative stage of the life cycle occurred not only on cells, but upon the individual. Each potential innovation in the vegetative phase passed through a new selective filter. Innova-

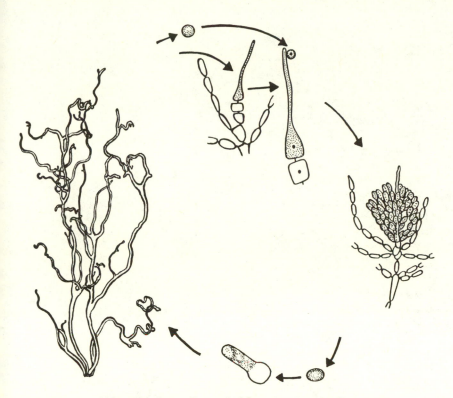

FIGURE 4.10　Life cycle of *Nemalion multifidum*, showing the diminutive diploid stage (shaded) borne atop the gametophyte. Details in text. (After Taylor, 1937 and Bold, 1967.)

tions which benefited both the cell and the individual were readily adopted, while those which created a conflict between the differing units of selection were restricted to those life cycles capable of resolving the conflict. The evolution of cellular differentiation is a cardinal case in point. Evolution of this trait was opposed at the level of the cell lineage, and only sexual organisms could effectively overcome this opposition. The first of many restrictions on the diversity of life cycles was established. Subsequent restrictions were ordained by the very different paths which individuality adopted in the three multicellular kingdoms. The ancestral characters that each clade brought to this fundamental evolutionary transition mediated, to greater or lesser extents, interactions between cell lineages within individuals and,

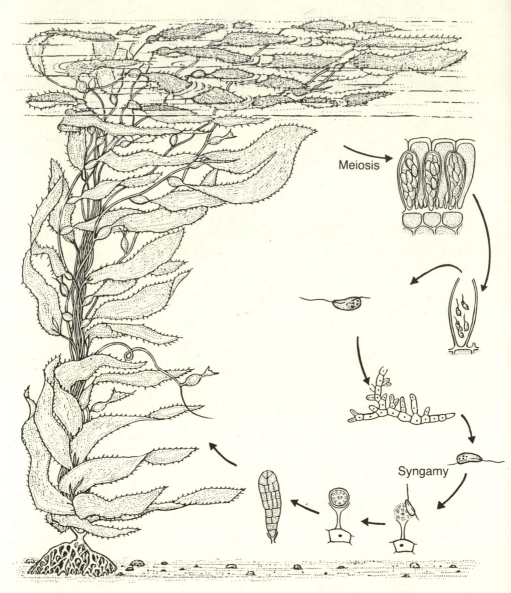

Meiosis

Syngamy

FIGURE 4.11   Life cycle of *Nereocytis luetkeana*, showing the enormous sporophyte arising from a diminutive haploid stage. Details in text. (From Bold, 1967.)

hence, determined the nature of potential synergisms and conflicts between selection at the level of the cell and of the individual. These differing constraints, in turn, framed what was possible in subsequent evolution, both creating additional limitations on life cycle diversity and providing opportunities for further life cycle modification.

Animal cells could move. Therefore their integrity could be threatened by somatic parasites, and their development became increasingly marked by early ontogenetic determination of cell fate. Early germ-line determination precluded asexuality, but fueled the evolution of bizarre life cycles among secondarily asexual forms. Cellularization demanded diploidy, and the susceptibility of metazoans to somatic parasites generated elaborate systems of historecognition. Fungi, starting from different conditions, followed a rather different progression. Lacking cells, their primitively haploid condition was preserved, but being coenocytic they were susceptible to deleterious variants. Variants became controlled by specialized patterns of cell-cycle synchronization, septation, and controlled cellularization. Fusion between individuals became a precisely controlled phenomenon, which in turn fueled the evolution of the heterokaryon as a unique mechanism of sexuality and formed the basis for elaborate pathogenic life cycles. Plants, with rigid cell walls, necessarily preserved an apical totipotent lineage and were less susceptible to somatic parasites. Having escaped the necessity of germ-line predetermination, plants could preserve asexuality and were largely immune from the detrimental effects of fusion. Yet, these same rigid walls removed the advantage of diploid cells over haploid counterparts in the somatic environment and allowed the development of complex alternating phases of each.

Individuality, in reflecting both the evolutionary resolution of conflicts between selection on the cell and on the individual and the evolutionary exploitation of synergisms between the two units, decreed that only a subset of all possible life cycles could actually occur, and that this subset should mirror the ancestral state of the somatic environment in which these conflicts and synergisms first arose. The interaction between the somatic ecology of the cell and the extrasomatic ecology of the individual set boundary conditions on life cycle diversification. The seven sets of traits discussed here (Table 4.3) can be combined in 972 possible combinations. Given the generalizations derived above, the diversity falls from the range of 10^3 to a mere 27 possible life cycles (Table 4.4). Modern life cycles are demonstrably shaped by the interplay of selection on those units of organization which alternate within them.

TABLE 4.3   Life Cycle Traits

---

1. CELL ARCHITECTURE
1.1 Coenocytic
1.2 Cellularized with rigid walls
1.3 Cellularized with non-rigid walls

2. SEXUALITY
2.1 Sexual
2.2 Asexual

3. ASEXUALITY
3.1 Primitively asexual
3.2 Secondarily asexual
3.3 Absent

4. DIVISION OF LABOR
4.1 Unicellular
4.2 Multicellular lacking cellular differentiation
4.3 Multicellular with cellular differentiation

5. GERM-LINE SEQUESTRATION
5.1 Present
5.2 Absent

6. HISTORECOGNITION SYSTEM
6.1 Fuse if similar
6.2 Fuse if different
6.3 Absent

7. PLOIDY
7.1 Haploid vegetative phase
7.2 Diploid vegetative phase
7.3 Alternating

TABLE 4.4

Restriction on Life Cycle Evolution	Number of Possible Life Cycles
1. No restriction	972
2. Sex is a necessary precondition to cellular differentiation	810
3. Non-rigid cell walls allow germ-line sequestration	540
4. Germ-line sequestration precludes primitive asexuality	486
5. Non-rigid cell walls or coenocytic condition require historecognition	319
6. Coenocytic condition invites fusion as a device for mate location	274
7. Vegetative diploidy is decreed by non-rigid cell walls	178
8. Vegetative haploidy preserved by coenocytic condition	148
9. Germ-line sequestration permits secondary asexuality	78
10. Biologically trivial cases[1]	27

1. Referring to traits listed in Table 4.3, the trivial cases are: if 1.2 or 1.3, not 4.1; if 2.2, not 3.3; if 4.1 or 4.2, not 5.1.

THE EVOLUTION OF HIERARCHICAL ORGANIZATION

Life is older than organisms. —G. EHRENSVARD, 1962

Summary

The history of life is a history of transitions between different units of selection. The transition focused on here—that between the cell and the individual—is but one such transition. Others must have preceded it. Numbered among those preceding must be the origin of self-replicating molecules, the association of autonomously replicating molecules into self-replicating complexes, the incorporation of such complexes into cells, the establishment of a multigenomic cell via incorporation of autonomously replicating organelles, and, with the evolution of sexuality, the origin of species. Just as transitions preceded the transition which established the multicellular individual, transitions establishing new units of selection have followed. Among these must be the association of individuals into kin groups and the association of neuronal activities into ideas capable of replication and variation.

When a transition occurs in the units of selection, synergisms between the higher and lower unit act to create new organizations which may allow the higher unit to interact effectively in the external environment. However, the organization of the higher unit does not simply interact with the external environment, it is also the agent of selection on the lower unit. To the extent that control over replication of the lower unit is required for effective interaction with the external environment, organizations must appear in the higher unit to limit the origin or expression of variation at the lower unit. Any such organization will act to stabilize the higher unit, as it limits the capacity for variants to arise or be expressed. Such processes are well known, although not emphasized in this context.

The organization of units of selection arising from ancient transitions bears these two features in common. First, each lower unit is selected by features of the organization of the higher unit: RNA-splicing may modify the expression of gene sequences independent of their rates of replication within the genome; the replication of organelles within eukaryotic cells is controlled by sequences on other organelles; and the fate of dividing cells is determined by patterns of somatic differentiation. In each case the lower unit may replicate, but only within bounds set by the influence of this replication on the higher unit's effectiveness in its interactions with its external envi-

ronment. Second, the unit itself is stable. The genetic code is universal, the intracellular community of organelles comprising eukaryotes is conserved at the kingdom level, and the fundamental bauplans of invertebrates are the same as those found a half billion years ago. Each unit has manifestly resisted significant perturbation over geological time scales. Inherent in any transition between two units of organization, therefore, is both the elaboration of a new organization and the stabilization of that same organization.

Life is hierarchically organized because any given unit of selection, once established, can come to follow the same progression of elaboration of a yet higher organization, followed by stabilization of the novel organization. Since this progression requires stabilization of a unit once established, it follows that the major features of evolution are established at times of transition between units of selection. The synthetic theory of evolution, which concentrates upon the individual as the primary unit of selection, is not capable of providing insights into the organization of those units preceding the evolution of the individual, that is, the organization of the genome, the organization of the cell, the organization of the eukaryotic community of organelles, or the organization of the developing embryo. Recognition that these features were shaped by the conflicts and synergisms between units of selection, particularly at times of transition, opens to evolutionary theory the ever-growing database of comparative developmental biology, cell biology, and molecular biology.

It is neither bold, nor even particularly insightful, to predict that this century will end as did the last—in an attempt to frame a consistent theoretical superstructure that merges development, genetics, and evolution. The close of the nineteenth century brought to an end an era of speculation on heredity, on development, and on phylogeny. Gone were the inheritance theories of Weismann and the biogenic law of Haeckel, displaced by the empirical tradition taking hold in the new science of genetics and Roux's experimental approach to embryology. Nearly a century of empiricism has now passed. The genetic material is known and may now be manipulated seemingly at will; the riddles of development appear within our technological grasp; and evolutionists stand ready, if not united, to attempt to incorporate the new-found complexity into Darwin's ever-robust theory.

August Weismann turned the century summarizing the

reflections of a lifetime in *The Evolution Theory*. His preface to this collection of lectures reveals a most curious irony. Weismann, in defining his "legacy to succeeding generations," wrote,

> In many details I may have made mistakes which the investigations of the future will correct, but as far as the basis of my theory is concerned I am confident: *the principle of selection does rule over all the categories of vital units.* It does not, indeed, create primary variations, but it determines the paths of evolution which these are to follow, and thus controls all differentiation, all ascent of organization, and ultimately the whole course of organic evolution on the earth. . . . This idea will endure even if everything else in the book should prove transient. [Italics in the original.][1]

Weismann's legacy to modern evolutionary biology could hardly be at greater odds with that which he had envisioned. He saw his main contribution in terms of selection on all "vital units,"[2] but this important point was lost to view. Other workers used his earlier germ-plasm theories as a basis for the modern emphasis on the individual as the primary, if not sole, unit of selection.

A century later, as evolutionists are in the midst of another attempt to synthesize the various biological specialties, a resurrection of Weismann's final conclusion may prove crucial. While genetics and development promise to contribute to the coming merger on the basis of new empirical discoveries, no one can seriously expect that the essential elements of the Darwinian notion of evolution will be fun-

1. Weismann, A. *The Evolution Theory*. Vol. 1. J. A. Thomson and M. R. Thomson, trans. London: Edward Arnold, 1904:viii–ix.

2. Weismann came to his view of selection acting on multiple units of biological organization late in his life under the influence of Roux (*Der Kampf der Teile im Organismus*. Leipzig: W. Engelmann, 1881). Roux's "selection of the parts" explicitly advocated a view of evolution based upon selection actively occurring not only upon individuals, but also upon the cells and molecules comprising organisms. Weismann, in characteristic fashion, absorbed Roux's notions into his general theoretical scheme, modifying the ideas in the process and relabeling the altered scheme "germinal selection" (Weismann, A. *On Germinal Selection as a Source of Definite Variation*. 2nd ed. London: Religious and Science Library, 1902). While the details of both Roux's views and Weismann's "germinal selection" are framed in language largely inappropriate to modern biology, the central notion of a hierarchical basis for evolutionary theory emerges unscathed.

damentally altered. Rather, evolutionary biology will actively contribute to the coming synthesis with its sister sciences only through a shift in language and emphasis. The language of the modern synthesis, entrenched in its emphasis on the individual, will be expanded as it is called upon to accommodate the natural history of the cell and of the genome.

II

The history of ideas is paved by the constraints of language. How we communicate frames what we can communicate. How we communicate frames what we see as trivially true and what we see as inherently paradoxical. How we communicate frames, in subtle and insidious ways, our very concepts of knowledge and hence what we claim to know.[3] As scientists we are typically free to function successfully in blissful ignorance of such issues. We share a common language, acquired through a long apprenticeship and jealously safeguarded by the social strictures of peer review. It is rare indeed that a formative concept falls on such hard times that a new language must be developed to take the place of verbal conventions whose time has passed. But such a time is upon us.[4] Few would debate that the Modern Synthesis, built on

3. Michael Polanyi (*Personal Knowledge*. University of Chicago Press, 1958) forcefully and persuasively argues that all knowledge is strictly *personal*. Each individual develops a distinct personal view of what is true. In this view, epistemology is an artificial discipline, which attempts, in vain, to establish objective criteria of truth or falsehood. A particular concept is true if a large number of individuals agree that it is true. All knowledge is personal and subjective, but often appears universal because language provides a mechanism to homogenize personal knowledge, leading to widespread adoption of particular notions as truths. Likewise, language constrains the potential classes of future "truths" to those which can be effectively communicated in the context of present language and, ultimately, weds the very notion of "progress" to social mechanisms for the development of new language.

Who has not felt the force of the inherently personal nature of knowledge in surprise at the views expressed by a colleague in peer review, or the homogenizing effects of language in one's response to such a review? Science gives the impression of progress (in contrast to other rigidly organized social subgroups, e.g., organized religion) because it is based on a set of social conventions that allow for the continuous growth of language. In this context, peer review represents a social device to "test" personal knowledge relative to the community mean.

4. In the words of L. L. Whyte (*Internal Factors in Evolution*. New York:

the explicit assumption that all evolutionary change is attributable to variation among individuals, is shortly due for
a considerable theoretical expansion in scope. Likewise, few
would doubt that this expansion will, in some pivotal way,
focus on what has been called the "units of selection problem." What remains in doubt is the language that we will
use.

Two early candidates are already in evidence. The first was
born in an unobtrusive manner with George C. Williams'
influential *Adaptation and Natural Selection*. Williams, in de-
bunking a variety of group selectionist claims, added a fundamental wrinkle to the units of selection problem. He interpreted each and every adaptation not in terms of group
advantage, but in terms of genic advantage. Genes were selected, not groups. Williams' shift in emphasis from individuals to genes went almost unnoticed. His interpretation
has not only peacefully coexisted with the synthetic theory
for two decades, but has also been typically regarded as a
brilliant defense of it. Williams' genic selection, however,
has taken on a new-found importance. When genic selection
was contrasted with selection on populations, it drew little
attention, as most people mentally equated genic selection
with individual selection. However, with the publication of
Richard Dawkins' *The Selfish Gene*, genic selection was pit-
ted against individual selection. In Dawkins's work, the significance of Williams' seemingly subtle shift in emphasis
became focused and clearly associated with a fundamental
shift in the language of evolution. Dawkins has forcefully argued

> . . . that the fundamental unit of selection, and there-
> fore of self-interest, is not the species, nor the group,
> nor even, strictly speaking, the individual. It is the
> gene, the unit of heredity.[5]

George Braziller, 1965:89), "The present situation is marked by a paradox of a kind frequent in the history of science: the operation of internal
factors [multiple units of selection] in several special contexts is already
regarded as commonplace, particularly in private discussions, but the fact
that this contradicts an assumption of the synthetic theory of evolution is
neglected in the literature, this being for many an emotionally charged
issue." I thank John Tyler Bonner for bringing Whyte's book to my attention.

5. Dawkins, R. 1976. *The Selfish Gene*. Oxford University Press, 1976:12.

Dawkins' claim is not only that genes are selected, but that *only* genes are selected. All else—the other candidates as potential units of selection—are merely "vehicles" to transport immortal genes from generation to generation.[6]

That genes are the unit of heredity and that we are merely mortal "survival machines" can hardly be doubted. The language of gene selectionism, though, is not the only language available. The original language of Roux and Weismann, wherein multiple units of selection are recognized, is also available. Lewontin adopts this language, recognizing any unit of biological organization as a unit of selection if it replicates, if in replicating it generates heritable variation, and if selection distinguishes between the variants.[7] These conditions are necessary and sufficient; any unit bearing these characteristics can evolve. Organisms, in this view, are not only units of selection themselves, but are also composed of many other hierarchically arranged units of selection.[8]

6. What is offensive to many in the genic selection view is its corollary that nothing more may be learned from the study of "vehicles" than can be learned from the study of genes. Stephen Jay Gould salts old wounds, stating, "I think, in short, that the fascination generated by Dawkins' theory arises from some bad habits of Western scientific thought—from attitudes (pardon the jargon) that we call atomism, reductionism, and determinism" (*The Panda's Thumb*. New York: W. W. Norton, 1980:92). George C. Williams (1986. *Oxford Surv. Evol. Biol.* 2:1–28) responds to such criticism by noting, "The history of science shows the inevitability of widespread emotional dissatisfaction with reductionistic interpretations of the world. I suppose there are people who think that the theory of sexual selection robs the grosbeak's song of its music or that kin selection makes a mockery of brotherly love. There used to be those who thought that Newtonian optics destroyed the beauty of the rainbow." The relationship between the gene selectionist view and the reductionist tradition is further, if somewhat less passionately, explored by William Wimsatt (in T. Nickles, ed. *Scientific Discovery*. Dordrecht: Reidel, 1980:213–259).

7. Lewontin, R. C. 1970. *Ann. Rev. Ecol. Syst.* 1:1–18.

8. In recent years there has been widespread recognition that ideas may behave as units of selection. That ideas vary and may be differentially communicated as a function of this variation can hardly be doubted. The current state of our understanding of cultural evolution is analogous to that of natural selection in Darwin's time: that transmissible variation occurs is hardly doubted by those who have given serious thought to the topic, but the mechanism giving rise to this variation is unknown.

If the ability to conceive of (or assimilate) a particular notion requires the existence of a particular spatial-temporal pattern of neuronal connections in the brain (i.e., an engram), then variation between individuals in the production (or assimilation) of ideas reflects an underlying physiological state at the time of creation (or communication). Variation in the pro-

Stephen Jay Gould, in championing the hierarchical view, recognized the units of selection debate as essentially a matter of definition: gene selectionism redefines the "unit of variation" as the "unit of selection." The choice here between the gene selectionist view, that of a single unit of selection, or the hierarchical view, that of multiple units of selection, is a choice of language. The crossroads at which we have arrived is not a choice between a correct and an incorrect perspective; it is a choice of how we will communicate in the future. We could adopt either language, relegating the philosophical difficulties of one or another mode to footnotes in a central theorem.[9] The question at issue is which language will teach us more.

---

duction and/or communication of new ideas, however, will only arise if this underlying state is malleable. Indeed, it appears to be. Synaptic connections increase in efficiency with stimulation and deteriorate with disuse. Thus, a particular individual's extended stimulation of a particular portion of a neuronal network may ultimately create novel spatial-temporal pathways permitting a new idea. The failure of another individual to either generate the same notion—or to be capable of effectively understanding the notion—may ultimately reflect the lack of stimulation of the proper subset of neuronal net (and/or deterioration of synaptic connections comprising the proper subset) required for transmission of the notion in question. This simple model is superficially supported by the commonplace observation that ideas arise as "brainstorms," perhaps as a consequence of sufficiently enhanced synaptic efficiency suddenly allowing a synaptic trajectory previously precluded, or by the observation that knowledge frequently cannot be transferred to some individuals (e.g., ask any parent of an adolescent), presumably reflecting the absence of a synaptic network capable of assimilating the idea in question. Enhanced synaptic efficiency via stimulation, and synaptic deterioration via disuse may serve as a source of transmissible variation in ideas. "Neural network" simulations may provide a technology with which to explore such relationships. One would like to ask, for example, whether the computational capacity of a model neural network increases, or displays emergent properties, when synaptic efficiency (and/or connectivity) is made a function of synaptic use. Whether or not this particular perspective has neurophysiological merit, the fact remains that an evolutionary perspective on brain function, i.e., viewing ideas as units of selection, must ultimately specify, *as the principal concern*, mechanisms for variation in the production and transmission of ideas within and between individuals.

9. Philosophical difficulties with gene selectionism have already occasioned comments by philosophers (e.g., Elliott Sober. *The Nature of Selection*. Cambridge, Mass.: MIT Press, 1985). Philosophical difficulties, however, have rarely stood as impediments to scientific discourse. Philosophical issues are traditionally shunted aside until the moment in history when the contradiction in question renders it impossible to incorporate some particularly glaring empirical data into the existing framework. I

George C. Williams makes precisely this point in the context of his criticism of various group selectionist schemes:

> Like the theory of genic selection, the theory of group selection is logically a tautology and there can be no sane doubt about the reality of the process. Rational criticism must center on the importance of the process and on its adequacy in explaining the phenomena attributed to it.[10]

It may seem foolhardy, in today's climate, to advocate anything but the genic selection view. The lion's share of research funds is directed at molecular approaches to biological problems. We cannot but expect that discoveries arising from basic research in the near future will occur primarily in the realm of gene structure and expression. The same will surely be true of applied sciences, as agriculturalists manipulate the genomic structure of crops in attempts to increase yield, buttress nutritional content, and erect shields from disease. Medical sciences, in particular, must continue to concentrate on this arena, as clinicians are still unable to cure most viral diseases (viruses are organisms that invite comparison with Dawkins' primitive "replicators"). Even in the domains of classical phylogeny and embryology, one may

---

ignore the philosophical objections to gene selectionism here, not because I find them bankrupt, but because I am confident that the outcome of this debate will not be decided on this basis. The choice of evolutionary language will surely be governed by that which proves the most fruitful, over the short term, in resolving the empirical issues with which most scientists grapple.

The choice of language has yet to be made precisely because proponents of *neither* the hierarchical nor the genic selectionist view have successfully utilized their new language to extend evolutionary discussion beyond its traditional subject matter. Richard Dawkins, for example, opens his book on gene selectionism (*The Extended Phenotype*. Oxford University Press, 1983:1) with the introduction, "This is a work of unabashed advocacy. . . . What I am advocating is not a new theory, not a hypothesis which can be verified, not a model that can be judged by its predictions. . . . Any reader who expected a convincing new theory in the conventional sense of the word is bound to be left, therefore, with a disappointed 'so what' feeling." Similarly, Niles Eldredge (*Unfinished Synthesis*. Oxford University Press, 1985:142), one of the more active proponents of hierarchical views, admits as much, saying "Much has been said about hierarchies in an evolutionary context in the past half-dozen years or so, yet relatively little concrete has been done about them."

10. Williams, G. C. *Adaptation and Natural Selection*. Princeton University Press, 1966:109.

easily anticipate discoveries which will surely fit comfortably with the genic selection view. It takes little insight to predict that, more likely sooner than later, some major case of evolutionary convergence will be found to have arisen by horizontal gene transfer or that somatic differentiation will be found to arise by genic rearrangement. Economic and political reality favors the language of genic selection.

In these essays, I have, nonetheless, adopted the opposite position: that of a hierarchical perspective on the units of selection problem. Indeed, my central thesis has been that the suite of adaptations that we refer to as development arose as a direct consequence of the transition from one unit of selection, that of the cell, to another, that of the individual. I need not have framed the arguments thus. Dawkins deals with these same issues in the language of genic selection in *The Extended Phenotype*, his sequel to *The Selfish Gene*. Under the heading, "Rediscovering the Organism," Dawkins admits:

> There really *is* something pretty impressive about organisms. If we could actually wear spectacles that made bodies transparent and displayed only DNA, the distribution of DNA that we would see in the world would be overwhelmingly non-random. If cell nuclei glowed like stars and all else was invisible, multicellular bodies would show up as close-packed galaxies with cavernous space between them. A million billion glowing pinpricks move in unison with each other and out of step with all the members of other such galaxies. [Italics in the original.][11]

How then does genic selection help us understand why such galaxies occur and why they develop in the peculiar and idiosyncratic ways that they do? Dawkins concludes:

> I have tried in this last chapter to sketch the general direction in which we might proceed in seeking an explanation. . . . I summarize it here. . . . The replicators that exist tend to be the ones that are good at manipulating the world to their own advantage. In doing this they exploit the opportunities offered by their environments, and an important aspect of the environment of a replicator is other replicators and their phenotypic

11. Dawkins, *op. cit.*, p. 250.

manifestations. Those replicators are successful whose beneficial phenotypic effects are conditional upon the presence of other replicators which happen to be common. The world therefore tends to become populated by mutually compatible sets of successful replicators, replicators that get on well together.[12]

In the genic selection perspective, life is organized into discrete individuals, and these discrete identities develop as they do because, for some as yet unspecified reason, this arrangement proved the most beneficial to the selfish interests of individual genes. If we adopt the gene selectionist view, the central evolutionary problem becomes that of describing how a given trait acts to serve the selfish interests of the gene.

Each of the arguments I have offered *could* have been phrased in this way. The origin of germ-line sequestration, for example, could have been phrased as the selfish pursuit by a replicator of its own individual advantage.[13] Germ-line sequestration is, I freely grant, little more than a selfish innovation.[14] However, in the gene selectionist view, the matter ends here; germ-line sequestration is simply one of a multitude of selfish adaptations. If, however, one admits both cells and individuals as units of selection, then germ-line sequestration is a particularly important selfish innovation, for it fundamentally alters the degree to which further variation

12. *Ibid.*, p. 264.
13. Gene selectionism is commonly framed as if genes alone meet the criterion of replicators. Dawkins defines replicators as "any entity in the universe of which copies are made" (*ibid.*, p. 293). In other words, replicators are clones. Not only genes are replicators; cells, organelles, and chromosomes qualify, as do asexually produced ramets. The gene selectionist perspective, therefore, could also be reframed in something of a hierarchical perspective, freeing it from its equating of the "unit of heritability" with the "unit of selection." I see little virtue in this exercise, however, as it would only produce a language parallel to the existing hierarchical perspective, but burdened by its failure to incorporate mixis as a form of reproduction. While in some specialized context this might well prove useful, it strikes me as an insufficient basis for so fundamental a revision.
14. While germ-line sequestration is a selfish innovation, it should perhaps be emphasized that it is a selfish act on the part of somatic cells, not the germ line. Germ-line sequestration arose through cells pursuing increased replication within the somatic environment. The germ line is a mitotically inactive lineage, a loser in cell lineage competition. The selfish cells here are the somatic cells, which abandoned a function of significance to the individual, in return for further replication.

expressed in a lower unit of selection can be inherited by a higher unit. Germ-line sequestration represents something more than a mere selfish act of a replicator—it represents the triumph of the level of the individual over the level of the cell.

The hierarchical view recognizes that both the cell and the individual are viable units of selection. Certain selfish adaptations were favored at both levels of organization, while others were favored at only one.[15] Germ-line sequestration represents a particular form of synergism between the two units of selection. This particular synergism, once established, acted to constrain the selfish pursuits of all future cells within organisms bearing this trait. Both the gene selectionist and the hierarchical view recognize germ-line sequestration as a selfish innovation. Only the hierarchical view, however, leads directly to the predictions that this particular selfish innovation stabilized the chaos of Precambrian

15. Stephen Jay Gould, in particular, has emphasized the role of synergisms and conflicts in the interpretation of the relative frequencies of biological phenomena. He has used the example of sexuality. Sex is said to be ubiquitous because sex acts as a mechanism of DNA repair (Bernstein, H., H. C. Byerly, F. A. Hopf, and R. E. Michod. 1985. *Science* 229:1277–1281). Sex is ubiquitous, according to the arguments I have presented, because it allowed the evolution of cellular differentiation. Sex is ubiquitous, according to Weismann (*Essays on Heredity and Kindred Biological Problems*), Williams (*Adaptation and Natural Selection*), and Maynard Smith (*The Evolution of Sex*), because it provides the individual with crucial genetic variation in the face of changing environments. Sex is ubiquitous because it offers to the group the new and advantageous mutations arising in just one of its members. Sex is ubiquitous because sex promotes rates of speciation and decreases rates of extinction (Stanley, S. M. 1976. *Science* 190:382–383). Gould, however, is certainly correct. Sex is ubiquitous for all these reasons. Sex is ubiquitous because it can be favored at the genic, the cellular, the individual, the populational, and the species level. John Tyler Bonner (*The Evolution of Complexity*. Princeton University Press, 1988) has recently developed a parallel explanation for evolutionary trends in size and histological complexity.

Perhaps the clearest example of the role of synergisms between different units of selection in guiding the evolution of organismal adaptation is found in the immune system of vertebrates. The immune response clearly favors the individual by permitting the successful repulsion of pathogens. At the cellular level, the immune response arises as a consequence of "selfish" behavior on the part of those cell lineages participating in the response. Finally, at the level of the gene, the diversity of receptors allowing the cellular response to occur arises from a number of independent sequences for receptor subunits dispersed throughout the genome, which likely arose via the "selfish" proliferation of receptor sequences.

development into functional bauplans, established hetero-chrony as the principle mechanism of evolutionary change within established bauplans, limited the longevity of organ-isms displaying it, constrained the occurrence of asexuality in the life cycle of higher metazoans, and fueled the evolu-tion of complex life cycles in forms secondarily acquiring asexuality.

If, in the coming synthesis, evolutionary theory is to per-form more than a "book-keeping" function[16]—if the role of the evolutionist is to be something more than that of an ac-countant engaged in providing post-hoc summaries of how a given mechanism ensures selfish interests—then some un-derstanding must be found on which to base predictions. In those arenas of natural history which evolutionary theory has yet to illuminate—genomic structure, cellular organization, and ontogenetic pattern—prediction will come not through the mere specification of why an innovation is selfish, but through an understanding of why some selfish devices were pursued in the course of evolutionary history and some were not. The selfish interests of genes have manifestly produced "vehicles" in the forms of organelles, cells, individuals, and yet higher units. If evolution is to predict as well as describe, then selfish interests must be understood in the framework of the constraints and opportunities generated by the "vehi-cles" they have created. Each such unit defines, in ways we are only beginning to perceive, those classes of selfish pur-suits which are possible and those which are not. Clearly those selfish pursuits which favor the replication of both the gene and the vehicle are more likely to survive than those which favor the gene but not the vehicle. On this basis, ev-olutionary theory can contribute to the coming synthesis with predictions based on synergisms between genes and the various vehicles. In other words, prediction will come by adopting a hierarchical view based on multiple units of se-lection.

Thus, I have chosen in these essays to adopt a hierarchical view, not simply because it has clear priority in the notions of Roux and Weismann, but because I believe that this lan-

16. William Wimsatt (*op. cit.*, see above note 6) and George C. Williams (*op. cit.*, see above, note 6) have explored the virtues of genic selectionism as an evolutionary accounting book, from the perspectives of critique and support, respectively.

guage offers to teach us more.[17] Both the gene selectionist view and the hierarchical view are useful perspectives for the purposes of *identifying and describing* evolutionary innovations. However, identification and description is the easy part of evolutionary theory. The complementary problem of defining how a particular selfish innovation has channeled subsequent evolution, and using this knowledge to explain patterns previously unaccounted for, is the hard part. The gene selectionist perspective offers no particular guidance here. The hierarchical view does. To fail to acknowledge a multiplicity in the units of selection, and to concentrate solely on the selfishness of each evolutionary innovation, is to miss what might be learned from the study of potential synergisms and conflicts between different units. To adopt a gene selection perspective is not wrong. It simply does not help unravel the central dilemma of our science. Evolution is historical, and our most difficult task is not to identify evolutionary innovations, the individual selfish acts, but to identify their sequence and consequences.

III

That life is hierarchically organized, with species composed of populations, populations of individuals, individuals of cells, cells of organelles, organelles of genomes, genomes of chromosomes, and chromosomes of genes, is so obvious an observation that it is quite remarkable that we have no general explanation of why this is so. Life, beginning as self-replicating molecules, did not persist in this initial state. Rather, it sequentially elaborated "vehicles" in which the original heritable units became increasingly distanced from direct interaction with the external environment. Why is life hierarchical?

When a new unit of organization arises, a fundamental

---

17. David Hull (1980. *Ann. Rev. Ecol. Syst.* 11:311–332) comes to a similar conclusion, noting: "Evolutionary biologists are currently confronted by a . . . dilemma: If they insist on formulating evolutionary theory in terms of commonsense entities, the resulting laws are likely to remain extremely variable and complicated; if they want simple laws, equally applicable to all entities of a particular sort, they must abandon their traditional ontology. This reconceptualization of the evolutionary process is certainly counterintuitive; its only justification is the increased scope, consistency, and power that results."

change occurs in the form of selection. Whereas before the appearance of a new unit, selection is imposed by the external environment alone, after the unit arises, selection by the external environment acts only upon the higher unit.[18] Selection on the lower unit, previously under strict control of the external environment, now occurs within the "environment" of the higher unit. *Traits expressed in the higher unit now act as selective agents on the variation arising in the lower unit.* The organization of the higher unit is, however, a function of prior variation in the lower unit. Thus, the lower unit can influence the replication of the higher unit by modification of its organization to suit the lower unit, but only to the extent that replication of the lower unit does not disadvantage the higher unit in its interaction with the external environment.[19]

18. The external environment will act solely upon the higher unit *only* if the lower unit is physically contained within the higher unit, as in the case of genes within cells or cells within multicellular organisms. When the lower unit is not physically enclosed within the higher unit, e.g., organisms within species, the external environment may actively select *both* units.

19. Kin selection is a case in which genic advantage does not disfavor the higher unit in its interaction with the external environment, and so is permitted. However, in this case, the relevant higher unit is not the individual, but the "family group." As Wade has pointed out (Wade, M. J. 1979. *Am. Nat.* 113:399–417), kin selection teaches not of the importance of the gene as a unit of selection, but of the family group as a unit of selection above the level of the individual. Cases of altruism, interpreted in terms of inclusive fitness, do not violate a hierarchical perspective—they simply represent an instance in which the higher unit (i.e., the family group) is favored by virtue of a genic advantage in restricted ecological circumstances.

Kin selection, thus, represents an instance in which a higher unit, the family group, and a lower unit, the gene, have synergistic effects, despite the fact that these effects are in conflict with the interests of the individual. Just as cases of synergisms between different units of selection are responsible for the ubiquity of some traits (e.g., sexuality), the occurrence of restrictions on the distribution of traits often reflects the occurrence of conflicts between units of selection. Kin selection occurs in restricted ecological contexts because only in those contexts can the synergism between genic advantage and family-group advantage overcome the conflicts between these units and the individual.

The importance of kin selection in certain restricted ecological contexts is often held to be a principal reason for adhering to a genic selection view. For example, Dawkins (1978. *Zeitschrift für Tierpsychol.* 47:61–76) states, ". . . Hamilton did not go far enough. Paradoxically, the logical conclusion to his ideas should be the eventual abandonment of his central concept of inclusive fitness. We should also move toward giving up the term

Inherent in the transition between one unit of selection and another are two factors, one acting to create novel organizations and the other acting to restrain further modification. Variants will arise which favor the replication of the lower unit by creating organizations in the higher unit that favor the higher unit in its interaction with the external environment. This first factor, that of synergisms between the two units, will act to create novel organizations allowing exploitation of the external environment in ways that the lower unit alone could not accomplish. Presumably, those modifications which arise will be a complex, history-dependent function of the materials available for "evolutionary tinkering" at the time of the transition. A second factor, however, is equally important. Variants will also arise which disrupt the higher unit, that is, they will favor the lower unit at the expense of the higher unit. The rate and magnitude of such conflicts must be limited, or the higher unit will perish. If variants arise in the lower unit whose effect is to limit the occurrence or magnitude of subsequent variation, then the higher unit will eventually become resistant to further perturbation.[20] Conflicts will be minimized and the gains realized by prior synergisms will be preserved and protected. A decrease in the rate of subsequent variation, if sufficient in magnitude, will transform a system from a nonconservative state to a conservative, equilibrial state—one in which permanent perturbation from that state is discouraged. *Both the creation of history-dependent novelties through synergisms and the production of locally equilibrial states are factors inherent in the transition between any two units of selection.*

At a transition between one unit of selection and another, the higher unit will, then, change as a complex historical function of early variations, until such time as the rate of variation decreases, after which further change will be relatively rare and small in magnitude. The sequence of history-dependent phenomena shaping the organization of new units of selection until equilibrial arrangements are created, ren-

---

'kin selection' as well as group selection and individual selection. Instead of all of these we should substitute the single term 'replicator selection.' " Yet Hamilton did not fully embrace such an extreme conclusion, perhaps because kin selection is not at odds with a hierarchical view at all.

20. This is, of course, equivalent to stating that the rate and magnitude of variation suddenly decrease at some point following the transition between one unit of selection and another.

dering present states history-independent, provides a simple explanation for why life is hierarchically arranged. If each new unit becomes relatively immune to further modification, then further evolutionary innovation necessarily becomes concentrated at higher units.[21] Hierarchical organization arises because each newly created unit in the hierarchy can not only explore synergisms allowing exploration of environments closed to the lower unit, but the higher unit can become resistant to further modification. As the now higher unit becomes resistant to perturbations, it may itself act to effect modifications at a newly adopted higher level, and the process will continue to yield hierarchical organization as a result. Life, once established, will come to array itself in an ever-increasing wardrobe of "vehicles."[22]

If life became hierarchically organized in this fashion, then the lower the unit of selection, the more conservative it should be—for the simple reason that a lower unit must have become resistant to further change if it was to give rise to a new higher unit. That this is the case is well known. At the level of genomic structure, the genetic code is essentially universal in all organisms. At the level of nuclear-cyto-

21. At Darwin's strong urging, the notion of evolution by natural selection and the notion of progress have been divorced from the outset. This discussion, however, suggests a process by which life can sequentially adopt hierarchically arranged "vehicles." In the limited sense that life evolves hierarchically—and we cannot be confident in proclaiming the process to have run its course—progress is inherent in the evolutionary process.

22. This process of ever-increasing levels of hierarchical organization with time will reach a limit only if some process limits the potential for systems to become equilibrial. That some ecological communities fail to meet equilibrial criteria may ultimately be attributable to mass extinction events, which may act to reset ecological interactions before such a state is realized (Gould, S. J. 1985. *Paleobiology* 11:2–12; Jablonski, D. 1986. *Science* 231:129–133). While effectively immutable symbioses are known (e.g., lichens, gut flora), they constitute only a small fraction of the organisms comprising modern communities. It is unfortunate that the fossil record does not provide adequate resolution to test the notion that the frequency of obligate interspecific mutualisms increases with time following mass extinction.

The notion sets the mind to fancy. If earth—or some earthlike planet—did not occur so close to an errant asteroid zone, then these arguments would suggest that life would continue to develop hierarchical entities unabated. The next obvious step would be the development of ecological communities as self-replicating complexes—obligate symbiotic communities. Perhaps science-fiction writers, in conceiving extraterrestrial organisms as bizarre symbiotic complexes, are not so far from the truth.

plasmic interactions, the transcription-translation cascade is nearly universal. At the level of eukaryotic cell structure, the complex of self-replicating organelles is conserved at the kingdom level. At the level of the individual, the essential bauplans of metazoans have been fixed for a half billion years. Each such trait defines an equilibrium point.[23] Each has proved manifestly resistant to perturbations occurring over geological time scales.

History, the sequential operation of constraint and opportunity, is not the sole determinant of biological pattern. Not all present states require recourse to prior history. Some conditions are equilibrial: understanding of functional relationships alone implies understanding of present state. But the fact that life is hierarchical redirects attention to the effect of history on biological phenomena. The major features of genomic organization, of cell architecture, and of organismal ontogeny arose as the products of history-dependent variation at the time of the transition from one unit of selection to another. Units that persist today do so because variants which restrained further interaction between two units of selection arose and fixed the organization of the unit in question at a given state.[24] Further modifications have occurred,

23. The uniformity of these features has traditionally been attributed to their early evolution and monophyly. This explanation, while surely necessary, is not sufficient. The fact that a trait arose early cannot be reconciled with an observation of present-day ubiquity without the added condition that the state be equilibrial, that the state is resistant to perturbation.

24. R. C. Lewontin (*The Genetic Basis of Evolutionary Change.* New York: Columbia University Press, 1974:317–318) makes essentially this point in an attempt to frame a multilocus model for evolutionary change, noting that a linkage model with five loci has a dimensionality of 32 and 4.3×10^9 steady state conditions. He laments that "the introduction of linkage, while terribly interesting and exciting from a theoretical standpoint, made a predictive theory an absurdity." Despite the innumerable states conceivable for an entire genome if analyzed at the level of the gene, simple predictions can be made by a theory at the level of the chromosome. Lewontin again: "The equilibrial structure (we cannot yet tell about the dynamical structure) of complex genetic systems is a function only of the effect of homozygosity of whole chromosome segments and of genetic map length of these segments. . . . The significance of the transformation of our problem from a micro- to a macro-description does not only lie in what can be measured, but also in that it exists." Lewontin's final point is well taken. Potential gene-level analysis allows innumerable possibilities, but as the genes became organized into a higher unit, the chromosome, they necessarily adopted an equilibrial arrangement in the higher unit. Had they failed to do so, chromosomes would not exist.

but they are necessarily minor in magnitude relative to the changes which first established the organization.

The fact that life is hierarchical and that lower units are conservative demands the interpretation that each unit persisting in life's hierarchy today must have become resistant to further modification. Viewed in the context of selection on multiple units at times of transition, the reason for this conservatism is apparent. Synergisms between the units drove the elaboration of a higher unit and conflicts arising between units were minimized by adaptations limiting further variation. This conclusion has the fascinating and crucial corollary that *the major features of evolution were shaped during periods of transition between units of selection.*[25]

IV

Units of selection have arisen throughout the history of life. Life presumably began as self-replicating molecules. In this era selection was necessarily on molecules; individuals, populations, species, and higher taxa are strained concepts here. Self-replicating molecules later became associated in self-replicating complexes. In this subsequent era selection must have acted on the individual components, bounded only by that which threatened the complex, that is, bounded only by selection on the complex. Still later, self-replicating complexes of molecules became associated into cells, and later yet, into multicellular individuals, populations, and species. No theory can describe the evolutionary progress from molecule to cell to individual without subscribing to Weismann's precept—without attributing to each individual unit of biological organization the capacity to be selected. A central truth of biology—one that few would challenge, but, oddly, one which few of us chose to teach—is that *the history of life is a history of the primacy of differing units of selection.*

This series of essays has focused on one such transition in

25. The phrase "major features of evolution" used here should not be confused with its use by George Gaylord Simpson, whose influential book of the same title addresses repeated patterns in the fossil record. I use the phrase differently, believing the major features of evolution to include the organization of replicating molecules, the metabolic machinery of cells, and the principal developmental bauplans, as well as those macroevolutionary trends addressed by Simpson.

the units of selection, the transition occurring sometime in the Precambrian from single-celled organisms to multicellular individuals with division of labor. The prior state, that of free-living totipotent cells, and the subsequent state, that of cellular-differentiated individuals, both meet the trinity of criteria for evolution: they replicate, replication is occasionally heritably imperfect, and selection distinguishes between the variants. The transition was marked by a shift in how selection operated on the lower unit. In the prior state, cells were selected solely by the external environment. In the subsequent state, cells were selected in the somatic environment: some escaped terminal differentiation to produce new individuals; others fell to terminal differentiation. In pursuit of their own replication, variant cell lineages interacted with other lineages in the somatic environment to establish epigenetic programs which fortuitously benefited the individual in its interaction with the external environment. Individuality evolved as certain cell lines, behaving in their own selfish interest, displayed germ-line determination and/or maternal predestination. These two traits not only benefited the cell lineage which displayed them and the individual which harbored them, but also acted to limit the capacity for any further variants to arise, thus stabilizing the individual as a discrete entity relatively invulnerable to further modification.

The evolution of individuality is but one such transition. Before the evolution of the individual must have come the evolution of the eukaryotic cellular community, wherein the multiple genomes comprising eukaryotic cells represent a parallel problem of synergism and conflict. Variations arising within individual organelles faced an intracellular selective filter, with those variations which favored both the replication of the organelle and the cell being differentially favored. Potential for synergisms abounded in the early oceans, as Lynn Margulis has persuasively argued. With the gradual accumulation of oxygen in the atmosphere in the early history of our planet, mitochondria afforded the cellular complex access to new and increasingly important habitats.[26] Similarly, the later acquisition of chloroplasts afforded the potential for autotrophy to their hosts. As

26. Margulis, L. *Symbiosis in Cell Evolution*. San Francisco: Freeman, 1981.

prokaryotic cells became associated with one another within a common cytoplasm, however, conflicts must surely have occurred whenever the replication of the organelle was favored over the replication of the complex.[27]

The mere existence of benefits to association between prokaryotic cells, however, is not sufficient to explain why the transition to the eukaryotic condition yielded a stable unit of selection. For the eukaryotic cell to have become the stable unit that it is, mechanisms which limit conflict between the various organellar units must have been erected.[28] Presumably any genome within the cell that proved capable of regulating the supply of agents required for the replication of other organelles in that same cell would act to eliminate, or minimize, the rate at which further conflicts—or synergisms—would arise. Any such trait, once established, would act to stabilize the existing cellular community.[29]

27. Such conflicts are demonstrably common (Eberhard, W. G. 1980. *Q. Rev. Biol.* 55:231–249; Buss, L. W., in *Population Biology and Evolution of Clonal Organisms*, see Chapter 3, note 9). In forms with uniparental inheritance of mitochondria, for example, mtDNA often codes for products resulting in sterility of the sex that does not propagate the organelle and the overproduction of gametes of the sex that does. The fact that such conflicts are an ongoing phenomenon in extant organisms is underscored by the observation of "restorer alleles" in nDNA, which act to override the mtDNA-induced sex-ratio distortion.

28. The evolution of a stable cellular community not only involved the addition of new replicating units to a complex, but may also have involved the *loss* of units of selection in the course of evolutionary history. Organellar DNA is known to have been incorporated in nDNA in some modern organisms. Conceivably, oDNA-to-nDNA gene transfer could act to remove all outward traces of an original unit of selection. Indeed, Lynn Margulis, in arguing that cilia and flagella are derived from symbiotic spirochetes, suggests an example of this process.

29. The process of an extranuclear agent acting to stabilize nuclear division has been observed in laboratory culture (Garvey, E. P. and D. V. Santi. 1986. *Science* 233:535–540). A stable antibiotic-resistant strain of *Leishmania* was found to have arisen by virtue of extrachromosomal circular DNA. This element is capable of autonomous replication, carries antibiotic genes, and confers mitotic stability on the cells containing it. Extrachromosomal elements which are unable to confer mitotic stability on the cells containing them succumb, despite carrying the same antibiotic genes. The generality of this phenomenon, i.e., the requirement for sequences which stabilize the relative replication rates of co-occurring, but otherwise autonomous replicating units, is underscored by the observation that the stability of both plasmids and chromosomes also requires mitotic-stabilizing sequences (Jayaram, M., Y.-Y. Li, and J. R. Broach. 1983. *Cell* 34:95–104; Szostak, J. W. and E. Blackburn. 1984. *Ann. Rev.*

Just as patterns in metazoan ontogeny may be interpreted as the consequence of selection within the somatic environment, the natural history of eukaryotic cellular organization may be penetrable through consideration of intracellular selection. Cellular organization may be expected to reflect arrangements which disproportionately favor both the replication of the organelle and the cellular complex and, in particular, to involve traits which act to limit the origin and/or expression of variation affecting organellar replication. Phyletic patterns in the distribution and movement of cytoskeletal components, the timing of cell cycles, the regulation of organellar biosynthesis, and innumerable like components of the natural history of eukaryotic cell biology invite evolutionists to propose hypotheses to account for these patterns. A synthetic theory of eukaryotic cellular organization may ultimately be based on a thorough reconstruction of the conflicts and synergisms attendant in the transition from the single-genome cell to the modern, multigenomic, eukaryotic state.

Yet earlier transitions in the units of selection surely occurred. Self-replicating molecules, once free-living, became encased within cells,[30] with all the accompanying biosynthetic opportunities that a lipid prison allowed as a buffer from the vicissitudes of the external medium and the difficulties inherent in effective communication through it.[31]

Biochem. 53:163–194; Clarke, L. and J. Carbon. 1985. *Ann. Rev. Genet.* 19:29–55; Lusky, M. and M. R. Botchan. 1984. *Cell* 36:391–401).

30. How this may have occurred is, of course, unclear. Douglas R. Green has pointed out to me that as a self-replicating complex became large, it would require increasingly precise alignment for its replication. A complex which captured lipid molecules from the surrounding medium or which synthesized lipids might be capable of holding them in place by hydrophilic and hydrophobic sites on the complex. By conformational folding of the molecule, replication might then occur in precise sequence between a lipid sandwich. In this scheme, cells arose as a consequence of the complex becoming entrapped within the lipid layers it has previously utilized as an aid to replication.

31. The embryonic nature of the hierarchical perspective on evolution is nowhere more evident than in the fact that we do not know what suborganismal features were originally units of selection, nor do we know the sequence in which they arose. This point is particularly clear in the case of the putative transition between self-replicating molecules and cells. Freeman Dyson (*The Origins of Life*. Cambridge University Press, 1985), for example, has argued that cells and self-replicating molecules developed independently, only to have, at some point, become symbiotic; and A. G.

With the packaging of replicating molecules into cells, yet another selective filter was established. The potential for synergisms was great, as biochemical events requiring novel chemical environments, or, minimally, the retention of translation products which would otherwise have diffused away, could fuel both the replication of the complex and that of the simple cell. Conflicts, no doubt, also arose. For the first time, the fate of replicating molecules became inextricably bound to their translation products. Translation products which could not be metabolized or lost through the cell membrane no doubt threatened both the replicating complex and the cell containing them. Selection on the level of this simplest cell must have quickly restrained the "selfish" propensities of replicating molecules to limit the extent of variation in replication and to gain precise control over the translation process. Presumably any process which limited the origin or free expression of potentially devastating gene products must have been favored. Likewise, elaborate systems for the control of transcription and translation, for the packaging and intracellular transport of gene products, and for effective modes of communication through the cell membrane must have arisen.[32] Just as the evolution of the individual as a unit invites hypotheses regarding ontogenetic

Cairns-Smith (*Genetic Takeover and the Mineral Origins of Life*. Cambridge University Press, 1982) has argued that clay particles served as the original genetic material, only to have been later supplanted by nucleic acids.
32. RNA, whose new-found catalytic properties make it more than ever the prime candidate for the role of primitive replicant-complex, replicates at low fidelity. Copying of an RNA molecule proceeds without checking the correctness of each preceding base pair. Errors are necessarily frequent. Once entombed within a lipid layer, this process must surely have generated errors that were inimical to continued coexistence within the cell. However, the encasement of the replicant molecule within a cell offered a new opportunity to buffer the cell from variation arising within it. By "reverse" transcription, the original RNA could create a new source of genetic material—DNA—whose replication limited the rate at which variants arose. Replication could now be initiated by an RNA primer and henceforth proceed by checking the correctness of preceding base pairs. While RNA retained partial control over its offspring, it abandoned its once primary role of replicator and assumed its current position in the transcription-translation cascade. The transition from RNA as the genetic material to DNA as the genetic material could not but have occurred in a cellular context. Without a cell to hold together the products of reverse transcription, RNA could gain no benefit in creating DNA, and without the selection on the cell to link the fate of RNA to its products, the RNA would not have been so constrained to reduce its error rate.

pattern, and the evolution of the multigenomic eukarayotic cell invites hypotheses regarding organellar disposition, the evolution of the cell as a unit invites evolutionary hypotheses regarding the construction of the transcription-translation apparatus, the principal pathways of cellular metabolism, and the basic cytoplasmic architecture for compartmentalization and intracellular transport. Evolutionary interpretations of natural history at this level of organization can also be framed in terms of conflicts and synergisms between a self-replicating molecule and a cell.

Yet before the occurrence of cells, life may have taken the form of self-replicating complexes of molecules composed of smaller units, each capable of self-replication. The first self-replicating entities, probably small nucleotides capable of autonomous replication, perhaps constituted the first unit of selection.[33] If translation products or intermediates in the autocatalytic cycle of one replicator acted as a replicase or derepressor for another replicator, once-independent replicators may have become linked together.[34] Synergisms occur at

33. The transition from small replicators to replicating complexes was first explored by Manfred Eigen (Eigen, M. and P. Schuster. *The Hypercycle.* Berlin: Springer-Verlag, 1979). Eigen points out that as a self-replicating molecule increases in size, it will necessarily suffer a corresponding increase in the frequency of errors during self-replication. At some point, such errors would have left larger replicators at a disadvantage compared to the smaller ones. Yet evolution clearly did not stop with small replicants. Eigen's solution is "the hypercycle," whereby independent replicators whose autocatalytic properties complemented one another became linked to one another in self-replicating complexes.

34. The process of a replicative entity exploiting the machinery of another replicating entity to drive its own replication is not a process lost forever in the evolutionary past. RNA viruses today do much the same thing. When an RNA phage enters the host cell, it first utilizes the translation apparatus of the host to synthesize a protein subunit which, in association with host proteins, produces phage-specific RNA-replicase. This replicase then directs the proliferation of phage-RNA until the metabolic contents of the host cell are exhausted. A self-replicative entity is catalyzed by its host, just as Eigen (*ibid.*) has argued the first units of selection did in becoming associated into self-replicating complexes.

One of the major difficulties in imagining how cellular life evolved is that of understanding how a replicant complex, lacking a metabolic machinery entirely, could have continued replication encased within a cell (see F. Dyson, *op. cit.*). This problem may be partially resolved by suggesting that the life cycle of a primitive cellular organism involved an alternation of an inactive, encapsulated cellular stage with an acellular, free-living proliferative phase. Common-sense reasoning suggests the inherent plausibility of such a life cycle: a replicant complex entombed within a

this level as well. Environments lethal to individual replicators, due to, say, an environmental deficiency in concentration of a required replicase, may become habitable by the provision of the needed molecule by other replicators in the cycle. Conflicts are also possible. Individual replicators may come to adopt different positions in the cascade through the production of variants which permit the use of intermediates in another replicator's cycle to catalyze their own.[35] The growing discipline of comparative molecular biology similarly invites evolutionary hypotheses framed in terms of conflicts and synergisms between actually or potentially autonomous portions of the genome. The topological properties of the folding of such molecules, the arrangement of coding and noncoding regions, and the site-specificity of mobile elements must be numbered among those patterns which ultimately may be penetrable in terms of the transition from the first self-replicating molecule to the large self-replicating complex of fluid construction.[36]

The units of selection that evolutionists of the twentieth century most frequently debate—individuals, populations, and species—are only the latest in a long history.[37] That the century following Darwin should have directed its attention

---

lipid or protein coat might conceivably be cut off from environmental replicases required to initiate replication. Such an organism would likely be favored in an environment which was periodically inimical to a free-living replicant complex, as must have frequently been the case in primeval oceans. In this view, RNA viruses continue such a life cycle today, living in the only remaining environment on earth where replicases are freely available in the environment—that is, within living cells.

35. Indeed, simulation models of Eigen's hypercycle scheme have found that such selfish behavior on the part of independent replicators can act to destabilize a complex under a wide range of conditions (Neisert, U., D. Harnasch, and C. Bresch. 1981. *J. Mol. Evol.* 17:348–353).

36. The view of a replicator molecule with translational capacity prior to the adoption of DNA as the primary vehicle for information storage is a particularly powerful one. In this perspective, potentially autonomous replicators coded for individual protein domains, which ultimately became linked together in a complex by noncoding regions that could be extruded to act as enzymes, in a manner akin to Eigen's hypercycles. This view has the merit of seeing RNA splicing as a primitive feature (thus according nicely with the intron-exon arrangement of the archeobacterial genome) and views transposable elements as primitive "cheaters" parasitic on the hypercycle (see Reanney, D. C. 1979. *Nature* 277:597–600).

37. Conflicts and synergisms are also inherent in the notion of ideas acting as units of selection. The appearance of mechanisms for the transmission of ideas had unparallelled utility in the transmission of mutually beneficial

to these units and primarily to selection at the level of the individual is hardly surprising. The theory of evolution was originally framed largely in terms of individuals because individuals were the simplest entities for a nineteenth-century naturalist to work with and, presumably, to think about. To confuse this matter of expediency with sufficient evolutionary theory, however, would be a serious error. In previous eras, self-replicating molecules, complexes of such molecules, and cells were sequentially the units of selection acted upon by the external environment, and a theory of evolution must accommodate selection on these units and the transitions between these units.[38] Unlike our nineteenth-century predecessors, whose only natural-history database was that of organismal design, the biologist of today has available an ever-growing natural history of cells reflected in the diversity of development, a natural history of subcellular organization reflected in the diversity of cytoplasmic architecture

---

ideas. Conflicts, no doubt, also arose. Douglas R. Green has pointed out to me the importance of such "intellectual parasitism" (he cites as an example Tom Sawyer's ploy in enticing his companions to whitewash the fence for him). Ideas may spread which fail to benefit those individuals to whom they spread. Indeed, the initial development of language must certainly have been fueled by this potential for "intellectual parasitism," as individuals attempted to benefit by the spread of ideas without being parasitized by those ideas which acted to their detriment. Language, in this view, acts as both a mechanism for the spread of ideas and as a control on the spread of ideas. Within an existing language the potential for novel ideas—either beneficial or parasitic—is restricted, as all individuals share the common language and are constrained to communicate within its constructs.

38. This list of units of selection is only partial. For example, at the subcellular level, chromosomes are well known to behave as units of selection under a variety of conditions. At the suborganismal level, a strong case could be built for polyps or zooids as units of selection in clonal invertebrates and somites as units of selection in metameric organisms. At the level of cultural affairs, ideas are clearly selected units. Furthermore it is apparent not only that units may be lost in the course of evolution, but also that the addition of new units of selection has not always progressed via the terminal addition of units at a "higher" level. New units may arise within an established unit. (For example, species became a unit as soon as sex evolved, long before the evolution of the multicellular individual. Likewise, zooids and metameres likely arose only after the evolution of individuality.) Indeed, the greatest challenge facing a hierarchical perspective on evolution is to define those relationships differentially preserved when a transition occurs between two units of selection, and subsequently to use these criteria to discover the actual sequence of units that have occurred in evolutionary history.

and inter-organellar communication, and a natural history of genomic architecture reflected in the sequence and conformational states of nucleic acids. Just as evolutionary theory has proven successful in elucidating organismal adaptation by concentrating on selection upon individuals, an expanded evolutionary theory may be expected to elucidate the diversity of cellular and subcellular natural history by concentrating on lower units of selection and the transitions between them.

V

The theory of evolution has never proven a static construct. It has continuously shifted in emphasis, if not logical structure, in response to the demands of the times. As new empirical discoveries have highlighted previously unsuspected patterns in adaptation, or produced controversy regarding the efficacy of components of the evolutionary process in explaining particular patterns of adaptation, the theory has become progressively modified. The Modern Synthesis, the latest in this series of the modifications, was born of a need to incorporate organismal natural history with the findings of Mendelian genetics. Accordingly, the latest evolutionary synthesis has proven enormously successful in illuminating the statistical vagrancies of gene frequencies within populations and associated microevolutionary change. This same synthesis explicitly excluded from explanation the problems of development, and necessarily excluded the then-unsuspected complexity of subcellular architecture.

Once again evolutionary theory is faced with new challenges to its capacity to explain the diversity of life. These new challenges can be accommodated without modification of the logical structure of the argument, by an expansion in scope of the units of selection to which this logical structure is applied.[39] A clear program is available: history-dependent states must be unraveled through an understanding of con-

39. While the logical structure of Darwinism seems secure, this should not be taken to imply that Darwinism, when expanded to encompass hierarchical considerations, will not be found to possess a mathematical structure previously unsuspected. Formal expansion of the theory, in terms of transition from nonconservative to conservative states, may prove quite general.

straints imposed on a given unit of selection, and history-independent states must be specified by discovery of those processes which limit the expression of heritable variation on a given unit of selection. Such a program holds the potential to expand the scope of evolutionary prediction—to allow evolutionary arguments to provide precise predictions and hypotheses about why development proceeds as it does, why cells are constructed from the components that they are, and why replicating molecules display the sequence organization and conformational properties that they do. Each such arena is currently as much virgin ground for evolutionary theory as was organismal design in the days following Darwin's synthesis. Evolutionary theory has only begun to broach the issues which it will ultimately be called upon to explain.

AUTHOR INDEX AND REFERENCE KEY

Boldface indicates reference for first author on that page.

TAXONOMIC INDEX

Boldface refers to page on which a figure appears.

Library of Congress Cataloging-in-Publication Data

Buss, Leo W.
  The evolution of individuality.

  Includes index.
  1. Developmental biology. 2. Evolution. I. Title.
QH491.B87 1987   574.3   87-45514
ISBN 0-691-08468-8 (alk. paper)
ISBN 0-691-08469-6 (pbk.)